The Sneaky Book for Boys

THE SNEAKY BOOK FOR BOYS

How to perform sneaky magic tricks, escape a grasp, craft a compass, walk through a postcard, survive in the wilderness, and learn about sneaky animals and insects, sneaky escapes, and sneaky human feats

Cy Tymony

Andrews McMeel
Publishing, LLC

Kansas City • Sydney • London

Andrews McMeel Publishing, LLC
an Andrews McMeel Universal company
1130 Walnut Street, Kansas City, Missouri 64106

www.andrewsmcmeel.com

12 13 14 15 MLY 10 9 8 7 6

ISBN: 978-0-7407-7313-6

Library of Congress Control Number: 2007941350

attention: schools and businesses

Disclaimer

This book is for the entertainment and edification of its readers. While reasonable care has been exercised with respect to its accuracy, the publisher and the author assume no responsibility for errors or omissions in its content. Nor do we assume liability for any damages resulting from use of the information presented here.

This book contains references to electrical safety that *must* be observed. *Do not use AC power for any projects listed.* Do not place or store magnets near such magnetically sensitive media as videotapes, audiotapes, or computer disks.

Disparities in materials and design methods and the application of the components may cause results to vary from those shown here. The publisher and the author disclaim any liability for injury that may result from the use, proper or improper, of the information contained in this book. We do not guarantee that the information contained herein is complete, safe, or accurate, nor should it be considered a substitute for your good judgment and common sense.

Nothing in this book should be construed or interpreted to infringe on the rights of other persons or to violate criminal statutes. We urge you to obey all laws and respect all rights, including property rights, of others.

Contents

Part III
Sneaky Resourcefulness

Part IV
Sneaky Animals and Humans

Acknowledgments

I'd like to thank my agents, Sheree Bykofsky and Janet Rosen, for believing in my "sneaky uses" book concept from the start. Special thanks to Katie Anderson, my editor at Andrews McMeel Publishing, for her invaluable insights and support.

I'm also grateful to the following people who helped spread the word about the first three "sneaky uses" books: Ira Flatow, Gayle Anderson, Susan Casey, Mark Frauenfelder, Sandy Cohen, Katey Schwartz, Cherie Courtade, Mike Suan, John Schatzel, Melissa Gwynne, Steve Cochran, Christopher G. Selfridge, Timothy M. Blangger, Charles Bergquist, Phillip M. Torrone, Paul and Zan Dubin Scott, Dana Vinke, Cynthia Hansen, Charles Powell, Harmonie Tangonan, and Bruce Pasarow.

I'm thankful for project evaluation and testing assistance provided by Sybil Smith, Isaac English, and Bill Melzer.

And a special thanks to Helen Cooper, Clyde Tymony, George and Zola Wright, Ronald Mitchell, and to my mother, Cloise Shaw, for providing positive motivation, resources, and support for an early foundation in science, and a love of reading.

Introduction

If you're a boy, you're a born sneak. You love to know something that others do not and fool your friends with magic tricks and illusions. Yet, some of you do not receive science instruction that relates to the real world, and this might explain why many of you do not excel in science later in life. *The Sneaky Book for Boys* provides a way to learn the basic principles of science, improvisational resourcefulness, and have fun at the same time.

You don't have to be MacGyver to adapt unique gadgets, secure a room from intruders, or get the upper hand on aggressors. You can learn how to be a real-life improviser in minutes using nothing but a hodgepodge of items fate has put at your disposal. Sure, it never hurts to have the smarts of Einstein or the strength of Superman, but it's not necessary. *The Sneaky Book for Boys* is packed with science projects, sneaky product reuse applications, magic tricks, and seemingly impossible feats that will stump your friends. You will amaze your friends at a moment's notice with just paper and cardboard while demonstrating how to conserve our resources.

Created for lovers of sneaky tricks, gadgetry, and trivia, *The Sneaky Book for Boys* is an amazing assortment of over forty fabulous build-it-yourself projects, self-defense and survival strategies, product reuse ideas, and fun magic tricks. *The Sneaky Book for Boys* also highlights incredible stories about sneaky humans and critters that will amaze and inspire. After reading this book, you will revel in your newfound powers and glance around the room with a sly grin.

You can start your entry into sneaky tricks and resourcefulness here.

Part I

Sneaky Tricks

Here are some fun and foolproof tricks you can do anytime to amaze your friends. Each sneaky trick uses odds and ends you have with you or around the house. The best part? These illusions don't require a big production, special skills, or dexterity.

Want to know how to spin paper with your hands without touching it, turn a black-and-white image into color, or step through a postcard? Break seemingly unbreakable string without scissors? Detect counterfeit currency with a common household item? Perform other sneaky magic tricks? It's all here.

With a little preparation and a personalized delivery, you'll be ready at a moment's notice to perform these amazing feats.

Before you show off your new tricks, be sure to memorize the sneaky magician's rules:

- Practice until you can do the sneaky trick with your eyes closed.
- Don't reveal the secret behind the trick.
- Don't repeat an illusion to the same audience.

Detect Counterfeit Currency

Whether it's a hundred-dollar bill or a one, getting stuck with counterfeit money is a fear many of us have. In the United States in 2002, $43 million in fake currency was circulated. When counterfeit currency is seized, neither consumers nor companies are compensated for the loss. So what can we do about it? This project describes two methods to tell good currency from bad.

The first method is a careful visual inspection of the bill. Compare a suspect note with a genuine note of the same denomination and series. Look for the following telltale signs:

1. The paper on a genuine bill has tiny red and blue fibers embedded in it. Counterfeit bills may have a few red and blue lines on them but they are printed on the surface and are not really embedded in the paper.
2. The portrait and the sawtooth points on Federal Reserve and Treasury seals are distinct and clear on the real thing.
3. The edge lines of the border on a genuine bill are sharp and unbroken.
4. The serial number on a good bill is evenly spaced and printed with the same color as the Treasury seal.

The second way to verify paper currency is to test the ink. How can we do this in a sneaky way, at home or in the office? Easy: by using one important feature of the ink used on U.S. currency. A legitimate bill has iron particles in the ink that are attracted to a strong magnet. To verify a bill, obtain a very strong magnet or a rare-earth magnet. Rare-earth magnets are extremely strong for their small size.

They can be obtained from electronic parts stores and scientific supply outlets. See the Resources section at the back of this book.

You can also use small refrigerator magnets, connecting them end to end to create collectively a much stronger single magnet. See **Figure 1**.

What's Needed

- Dollar bill
- Strong magnet

What to Do

Fold the bill about in half crosswise and lay it on a table, as shown in **Figure 2**. Point the strong magnet near the portrait of the president on the bill, but do not touch it. A legitimate bill will move toward the magnet, as shown in **Figure 3**.

FIGURE 1

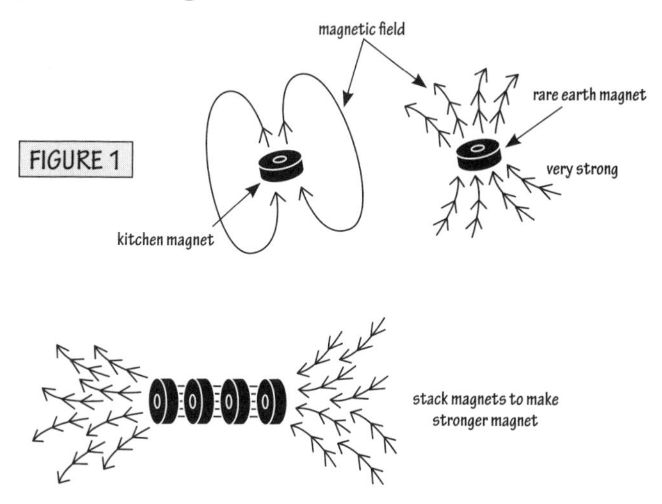

Whenever you doubt the authenticity of paper currency, simply pull out your magnet and perform the magnetic attraction test.

Sneaky Hiding Places

Hide and Sneak: Secure Valuables in Everyday Things

You've seen movies where a character hides something at home and you think, That's the first place I'd look! Well, this project will illustrate how to choose sneaky locations that are the last places a Man of Steal would look. You don't always have access to a safe deposit box or can install alarms on all of your possessions. But you can find sneaky hide-in-plain-sight places to frustrate and waste a thief's time.

Selecting this hiding place generally depends on two factors: the size of the item and the frequency of access required. From a package of soap to a tennis ball, a typical home offers a variety of clever hiding places, as shown in **Figure 1**. Wrapping your valuables in black plastic bags will further prevent discovery.

With enough time, a tenacious thief can eventually find virtually anything you hide. That's why you should have a room entry alarm installed in combination with sneaky hiding places to reduce the time a thief will spend searching for your valuables.

More Hide and Sneak

When you think of sneaky you usually think of something that is secret or hidden from you. Actually, the most common sneaky-use application is hiding your valuable belongings from others.

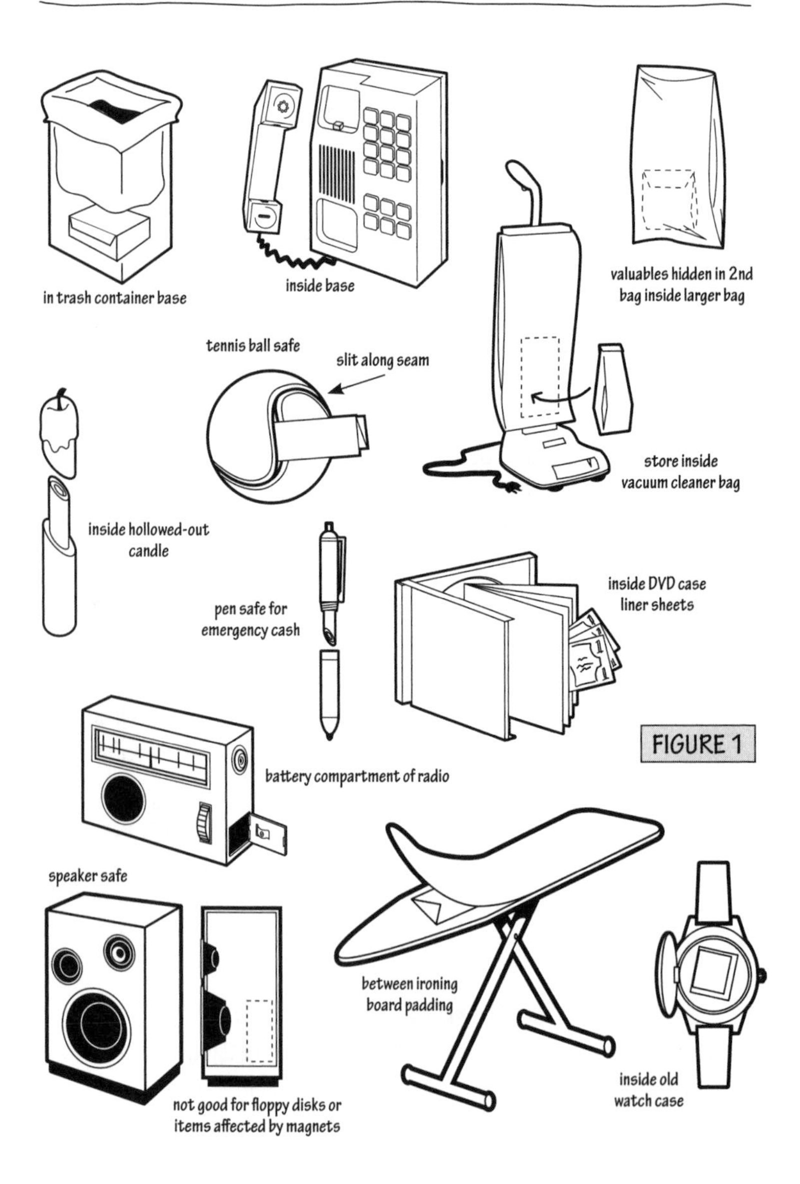
in trash container base
inside base
valuables hidden in 2nd bag inside larger bag
tennis ball safe
slit along seam
store inside vacuum cleaner bag
inside hollowed-out candle
pen safe for emergency cash
inside DVD case liner sheets
FIGURE 1
battery compartment of radio
speaker safe
between ironing board padding
not good for floppy disks or items affected by magnets
inside old watch case

The following ideas can be used to keep your things to yourself, even if they are in plain sight. Most likely a thief or nosy houseguest will briefly examine the item and then ignore it as a possible safe. See the list below for more examples.

- Figurine
- Tissue container
- Video or audio cassette shell
- Inner pocket
- Shoestring
- Between magazine pages
- Inside a candy box

How to Escape a Grasp

When possible, avoid physical confrontation. If necessary, call for help. Throwing punches is futile and dangerous, and in many cases, you injure your hand when you hit the other person. But in some instances, you may be attacked or grabbed without warning. In these situations, you must do something. The following techniques show easy-to-learn escapes that do not require lots of strength and can be quickly mastered.

As shown in **Figure 1**, the weak spot on the hand is between the thumb and forefinger. When an assailant grabs you, you will have a big advantage toward escaping his grasp if you locate this weak spot on his hand.

By pulling or twisting against the weak spot, as shown in **Figures 2** and **3**, you will escape from an assailant's grasp without a lot of effort. Practice this technique with a friend until it's a reflex action.

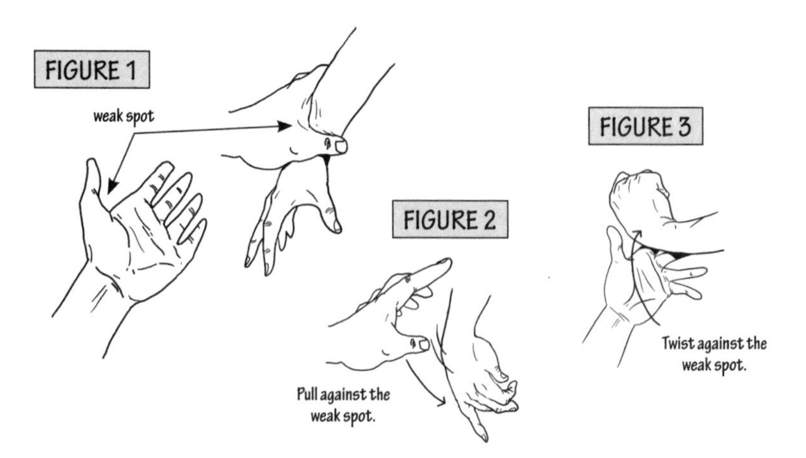

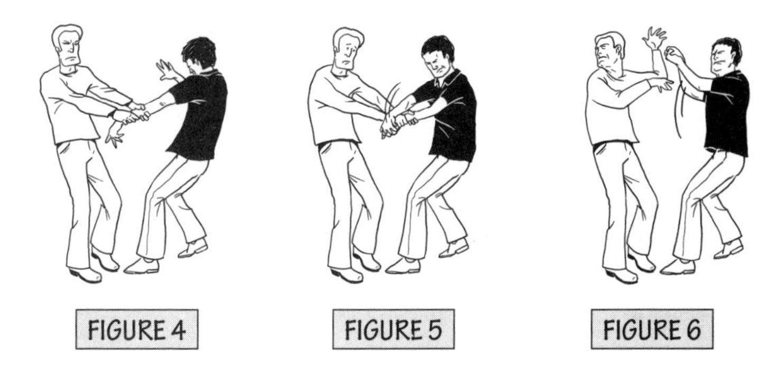

FIGURE 4 FIGURE 5 FIGURE 6

If someone grabs your arm with both hands as shown in **Figure 4**, use your free arm to grab your other hand as shown in **Figure 5**. Then raise both arms and turn (**Figure 6**). This will affect the weak spot on both of the assailant's hands causing him to release his grasp.

Similarly, if an assailant grabs both of your arms, as shown in **Figure 7**, rapidly pull your arms upward to force him to release his grip. See **Figure 8**.

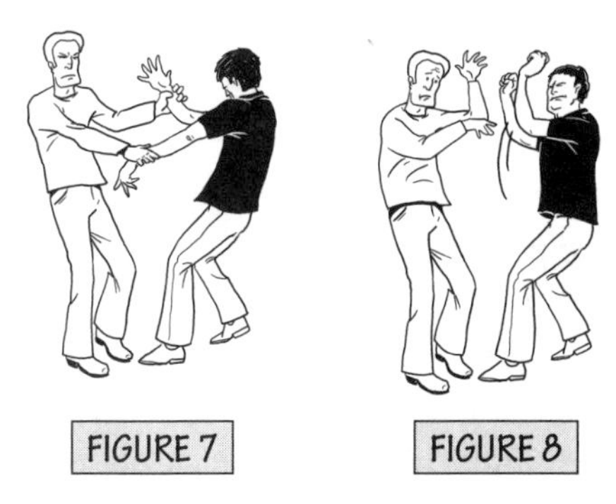

FIGURE 7 FIGURE 8

Sneaky Tracing Machine

Some items cannot be copied using a chemical transfer technique because their images are printed on coated paper or, in the case of text, the image will be reversed.

Another way to make a duplicate of an original image is to use a Sneaky Tracer.

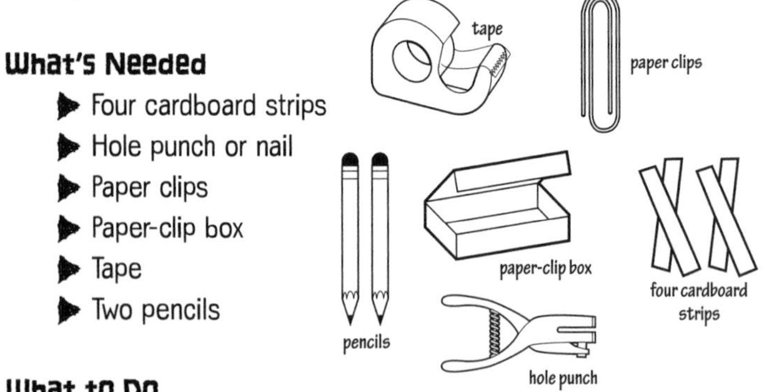

What's Needed

- Four cardboard strips
- Hole punch or nail
- Paper clips
- Paper-clip box
- Tape
- Two pencils

What to Do

First, cut two pieces of cardboard, each measuring 2 by 8 inches. Then cut another two pieces, each 2 by 4 inches.

Arrange the cardboard pieces in the pattern shown in **Figure 1** and then punch holes in the corners of the shape. Bend paper clips into C shapes and push them through the holes to secure the pieces, yet still allow them to move freely.

Next, punch pencil-sized holes in the cardboard at points A and B, just large enough so a pencil can fit through snugly, and insert a pencil in each (see **Figure 2**). Now place the copier device so one end rests on the top of the paper-clip box and secure it with tape.

The box acts as an elevated mounting platform to keep the pencils balanced and stable yet free to move about.

Last, select an original drawing that you want to trace and set it under the pencil in hole A. Place a blank sheet of paper under hole B. Use pencil A to trace the drawing and you'll see another picture being created by pencil B. If necessary, secure the pencils to the cardboard and the paper to the table with tape. See **Figure 3**.

Now you can easily trace complex drawings and make copies for your needs. Experiment with the lengths of cardboard, and you'll see that you can easily enlarge or reduce the size of the drawings made.

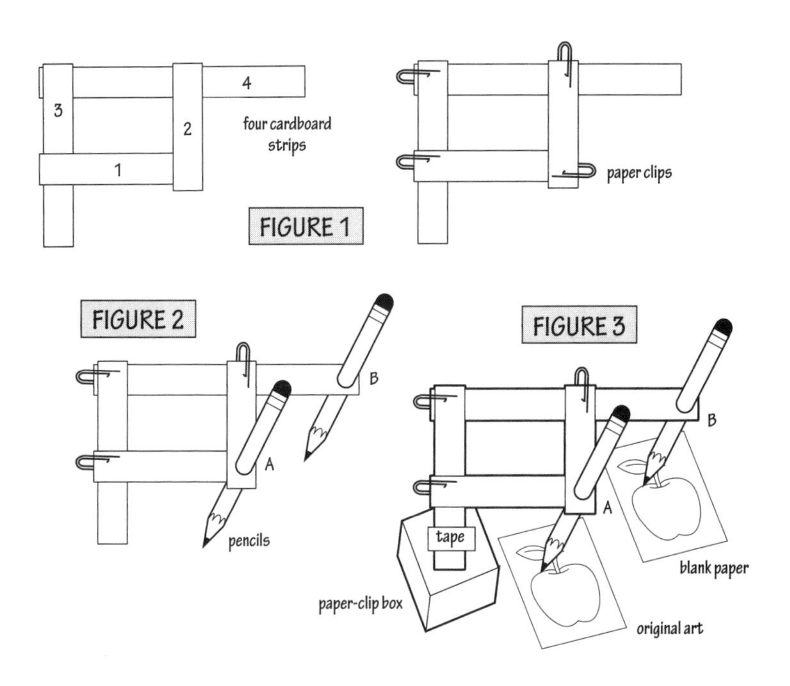

Sneaky Magic Tricks

Here are some fun and useful tricks you can do anytime to amaze your friends. The best part? Each sneaky trick uses common objects you have with you or around the house.

Make a Coin Vanish in Your Hand

Making an object disappear before the eyes of an audience is one of the most popular magic tricks. With a little practice, you'll be able to perform this simple trick anywhere with any small object.

What's Needed

- Coin

What to Do

Hold the coin upright in full view between your thumb and forefinger as shown in **Figure 1**.

Note: You can perform the trick in either hand. In this example, the trick begins in the left hand.

Next, place the thumb and middle finger of your right hand through the thumb and forefinger of your left hand as if you're about to grab the coin. See **Figure 2**. As you cover the coin with your fingers, and obscure its view from onlookers, let the coin drop into the palm of your left hand as shown in **Figure 3**.

Don't immediately close your left hand. Instead, close your right hand into a fist, swing it away from your left hand, and bring it near your onlookers to bring attention to it. Show them that it's empty and point to your left hand and the coin.

To your onlookers, it appears that you grabbed the coin from your left hand and the object is in your right hand's closed fist. However, the coin was never removed from your left hand because you sneakily dropped it from your thumb and forefinger into your palm. You can also bring your left hand behind an onlooker's ear and say that you made it travel there while you open your hand to reveal the coin.

FIGURE 1

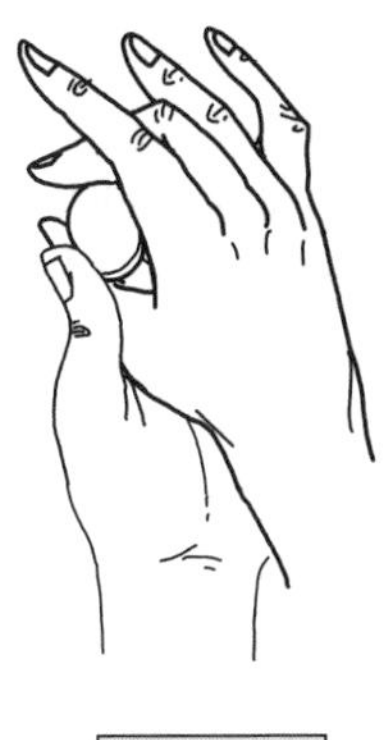

FIGURE 2

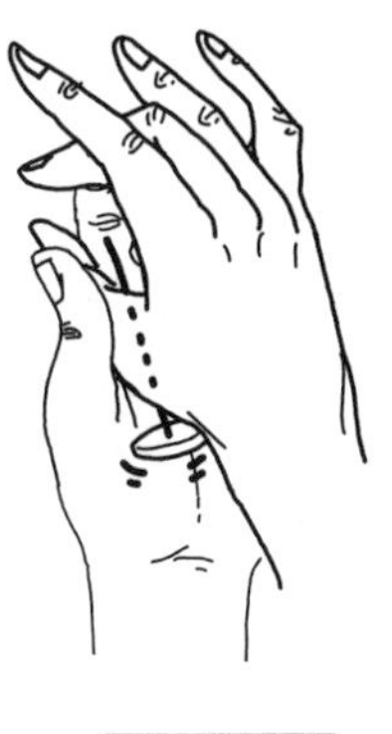

FIGURE 3

Balancing Soda Can

In this surprising trick, a soda can balances at a 45-degree angle on a surface or on your hand.

What's Needed

- Soda can
- Water

What to Do

First, pour out two-thirds of the soda from the can. See **Figure 1**. If you start with an empty can, pour water in it until it's one-third full.

Next, tilt the can on its edge so it stays in place as shown in **Figure 2**. If the can does not balance, add or remove liquid until the can stands on its edge.

Try balancing the can on a surface and even on your finger. For dramatic effect, act as if it takes lots of effort and skill to keep the can balanced.

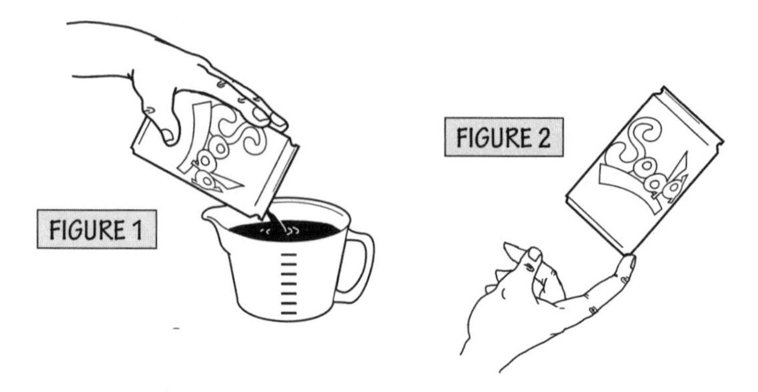
FIGURE 1
FIGURE 2

Tie a Knot With One Hand Without Letting Go of the String

Here's a sneaky dare you can make with a friend. Challenge your friend to pick up a length of string or rope and tie a knot using only one hand. After your friend's failed attempt, you'll execute this one-handed trick. You can also perform this trick with a tightly wound scarf.

What's Needed

- String, rope, or scarf

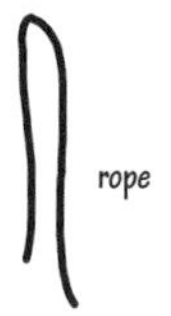

What to Do

First, place the string over your hand, but not your thumb, with one end hanging a few inches lower over the right side of your hand as shown in **Figure 1**.

Next, pull up the longer end using your middle and forefingers. See **Figure 2**.

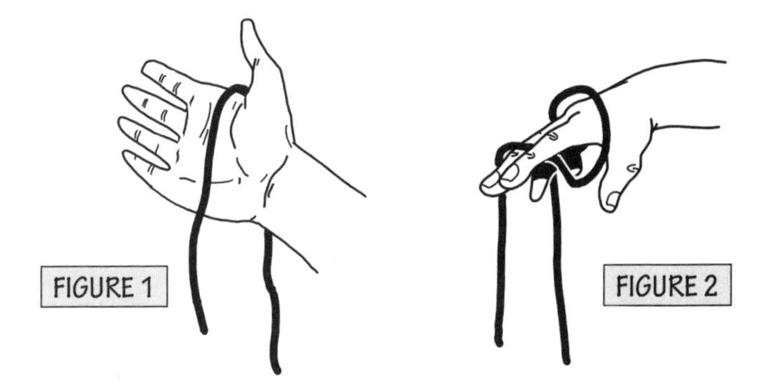

Wiggle your hand as you let the longer length of string pull through the short end, which now encircles it. Let the loop that has formed slip over the back of your hand as the long end pulls through the center. Keep wiggling your hand and soon the string will fall into a loose knot as shown in **Figure 3**.

Last, say "Presto!" and allow the string to drop down into a knot as shown in **Figure 4**.

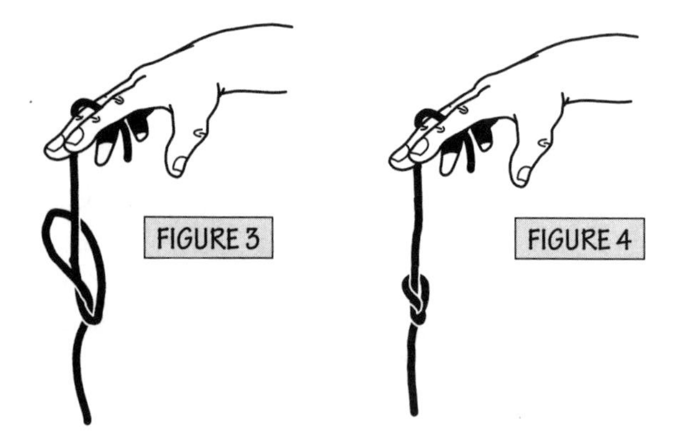

Tie a Knot With Two Hands Without Letting Go of the String

After amazing your friend with the one-handed knot-tying trick, dare him to pick up a length of string (rope or a tightly wound scarf will also work) with the forefingers and thumbs of both hands and tie a knot without letting go.

What's Needed

- Two-foot length of string, rope, or a rolled up scarf

rope

What to Do

First, place the string in front of you on a table, then cross your arms as shown in **Figure 1**.

Next, grab one end of the string with one hand and then position your arms so you can pick up the other end with your other hand. See **Figure 2**.

Last, uncross your arms and the string will magically tie a knot on its own as shown in **Figure 3**.

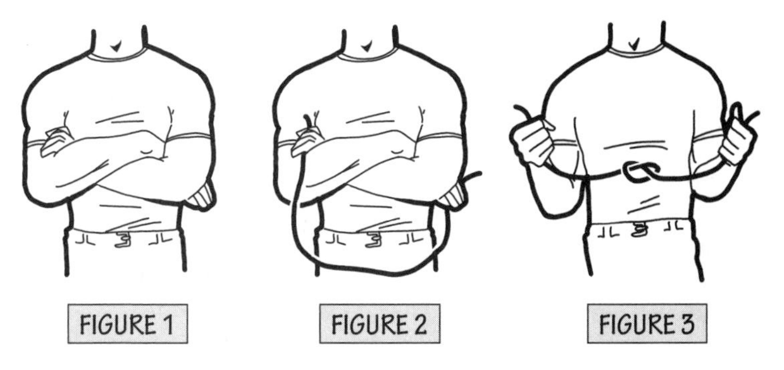

FIGURE 1 FIGURE 2 FIGURE 3

Step through a Postcard

You can win quite a few bets by performing this simple trick. Ask a friend to cut a hole in a postcard large enough to step through. Your friend can't cut strips and tape them together though.

When your friend fails to accomplish this feat, you can quickly make sneaky cuts in the postcard and amaze your onlookers.

What's Needed

- Postcard
- Scissors

What to Do

Figure 1 shows a typical postcard. Fold the postcard in half, widthwise.

Note: You can also perform this trick with a piece of paper, a business card, and so forth.

Using the scissors, cut along the lines shown in **Figure 2**. You can add more lines if desired but there must be an odd number of lines that alternate from the top and bottom of the postcard.

Next, unfold the postcard and cut across the three lines shown in **Figure 3.**

Last, pull the postcard apart along the cut edges until it forms a large circular shape large enough for you to step through. See **Figure 4.**

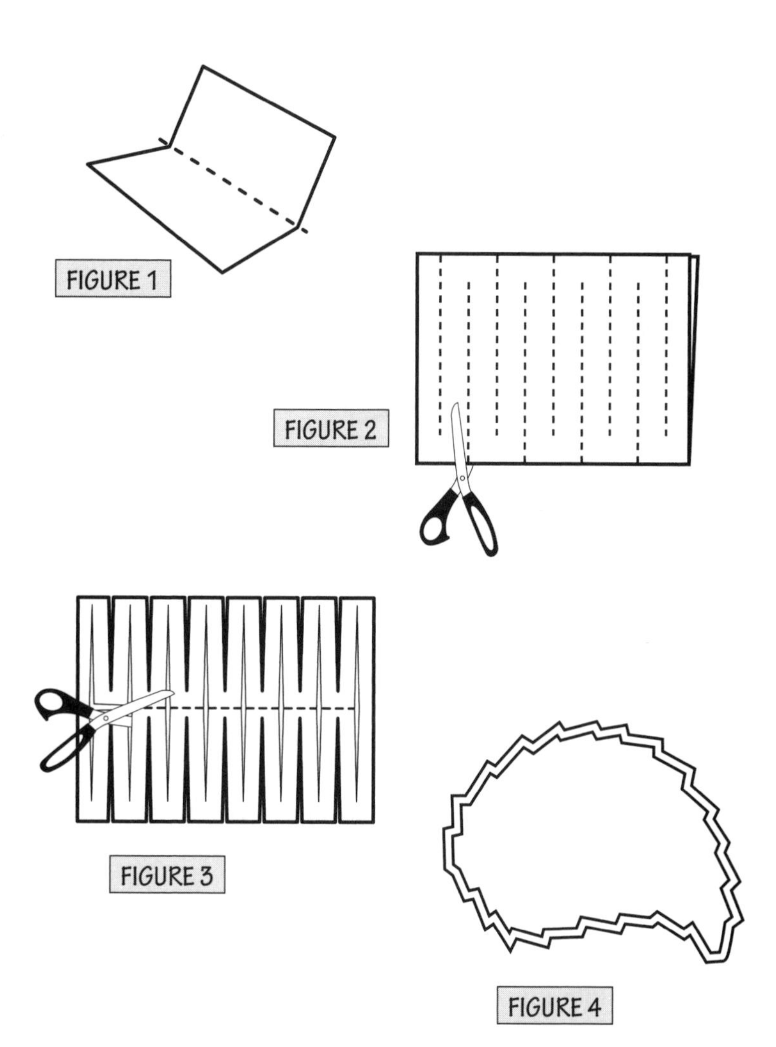
FIGURE 1
FIGURE 2
FIGURE 3
FIGURE 4

Sneaky Color from Black and White

Challenge your friends by asking them to change the color of an item without touching or painting it. Then amaze your friends when you actually produce colors from a black-and-white image with this next trick.

What's Needed

- White cardboard
- Black marker
- Pencil
- Pin
- Scissors

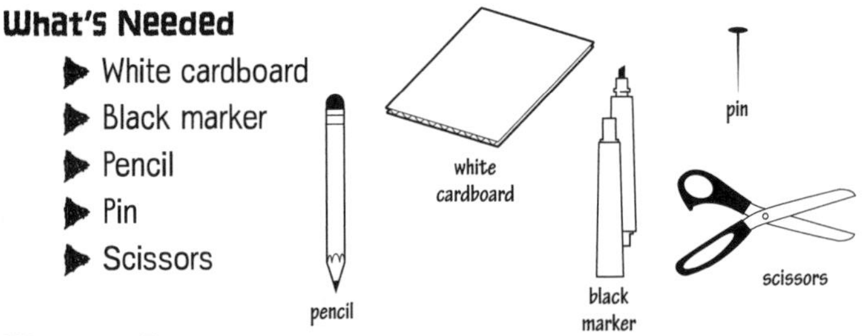

What to Do

Draw the disc shown in **Figure 1** on the white cardboard. Ensure that half the disc is solid black and half has the broken circle picture. If you photocopy the illustration, be sure to fill in blank spots with a marker. The disc should be approximately 4 inches in diameter. Cut out the disc with the scissors.

Place the disc on the center of the pencil eraser. Secure the disc to the eraser by pushing the pin into the eraser through the cardboard. See **Figure 2**.

Next, place the pencil between your palms and spin it. You'll see the black-and-white image turn blue and red depending on the speed, as shown in **Figure 3**.

How It Works

Your eyes have rod cells for peripheral, or side, vision, and cone cells for front vision and for discerning color.

There are three different types of cone cells that each respond to red, green, and blue at different response times. When you see white, all three cone cells respond equally. When you spin the disc, the alternating black sections, with their different lengths, cause an imbalance to how the three cone cells respond to white and your cone cells cause you to "see" different colors.

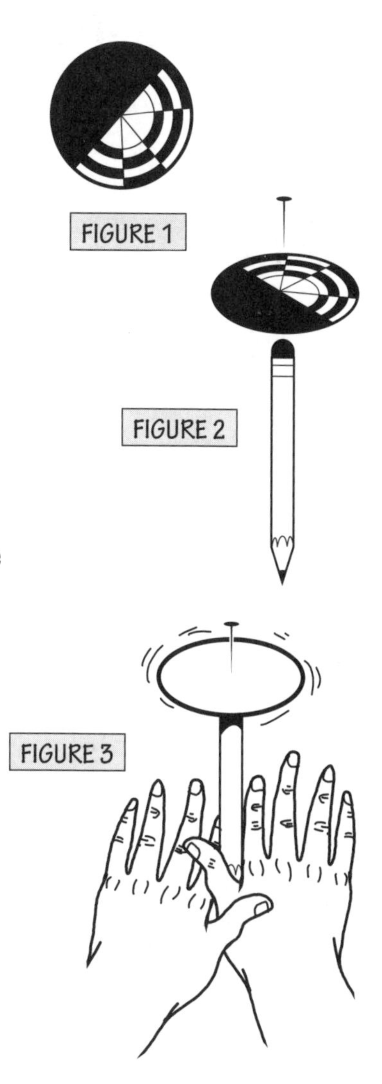

Hand-Powered Fan

As you may know, hot air rises. Rising heat can be made to move objects, and you can demonstrate this fact with a novel "hand-powered" motor. In this demonstrational science project, your hands will actually provide the heat to demonstrate how moving air currents can move an object in a rotary motion.

All it takes is an ordinary piece of paper, scissors, a needle, a cardboard box, and your hands.

What's Needed

- Paper
- Scissors
- Sewing needle
- Small cardboard box

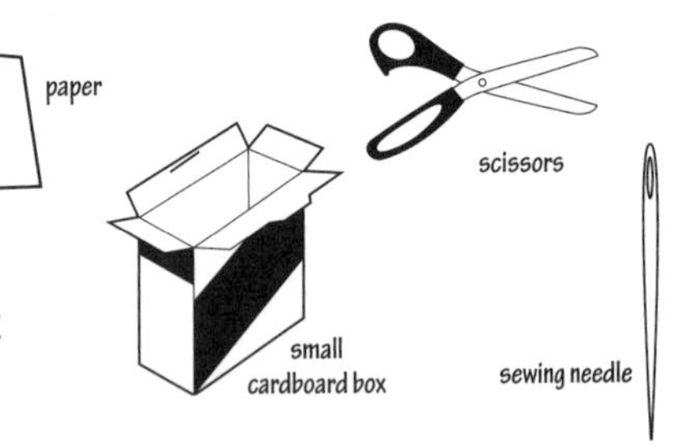

What to Do

Cut a piece of paper into a 2-inch square. Fold it in half diagonally; then unfold it and fold it in half on the other diagonal, as shown in **Figure 1**. This should create a cross-fold with a center point.

FIGURE 1

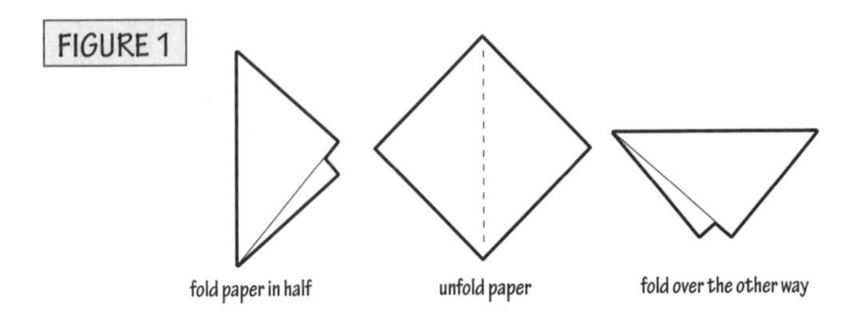

You can use a paper-clip box or similar small box as a mount for the needle. Hold the needle on its side with your fingers and carefully twist it into the top of the box (or use a thimble) until it punctures a hole in the top. Place the piece of paper on top of the needle so its center point allows the paper to turn freely. See **Figure 2**.

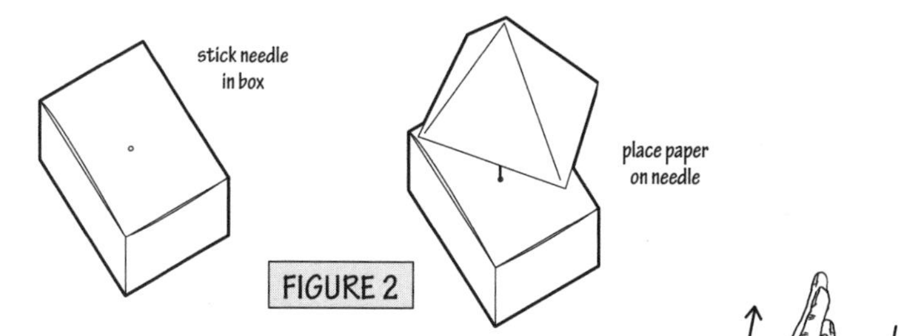

FIGURE 2

FIGURE 3

rub hands together

To make the sneaky "motor" turn, rub your hands together back and forth about twenty times to generate heat and place them near the sides of the paper. After a few seconds, the paper will begin to spin (**Figure 3**).

The paper spins because the heat on your hands causes a temperature increase in the air around the paper. As the heated air rises and cooler air takes its place, the air movement pushes the paper sides, causing it to rotate like a motor.

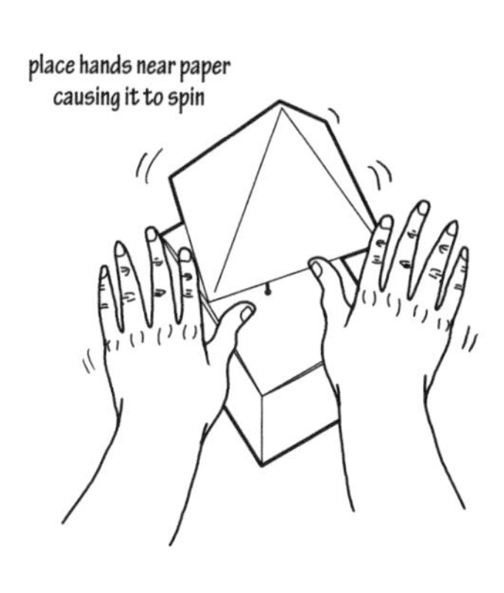

Sneaky Break String Without Scissors

Have you ever failed to break a length of string no matter how hard you tried? Want to look like a hero with super strength when others fail this test? This neat trick will also come in handy when scissors aren't around.

Note: This sneaky trick will work on thin cotton string, not with nylon or coated string.

What's Needed

- Cotton string, approximately three-feet long

What to Do

Wrap one end of the string three times around the forefinger of your left hand and then make a small loop near the palm.

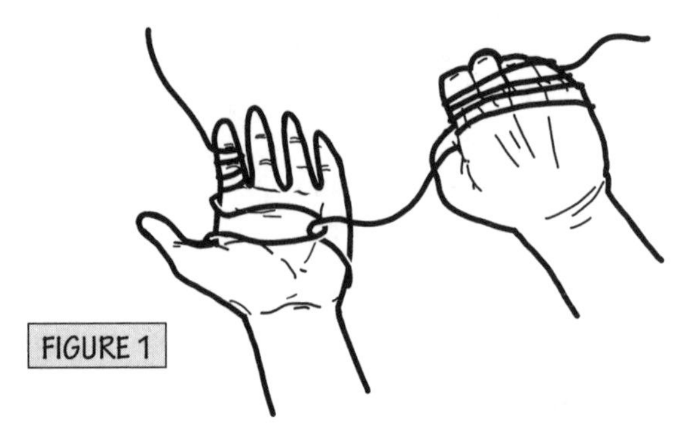

FIGURE 1

Next, pull the string behind your left hand and up and through the loop, and leave a 1-foot length between your hands.

Last, wrap the string at least four times around your right hand. To break the string, make a strong fist with both hands. Bring your fists together with your left hand on top, then rapidly drop your right hand. The string should break in the area of the loop. See **Figure 1.**

Note: First try this sneaky trick with thin string and then gradually try stronger string.

Levitating Art Figures

You can make everyday things balance in sneaky ways when you know the secret to determining the center of gravity. The center of gravity is the point in an object at which its mass is in equilibrium. Where this point is depends on the object's shape and weight distribution, and you can produce some attention-getting creations with this knowledge.

The following four projects are easy to do with items found just about everywhere.

Sneaky Balancer I

Knowing how to lower the center of gravity of an object allows you to produce figures that seemingly defy gravity (or make you seem like a skilled magician). This project demonstrates what happens when two similar cardboard figures have their center of gravity in different positions.

What's Needed

- Scissors
- Cardboard, a piece $8\frac{1}{2}$ by 11 inches

Optional:

- Sewing thread

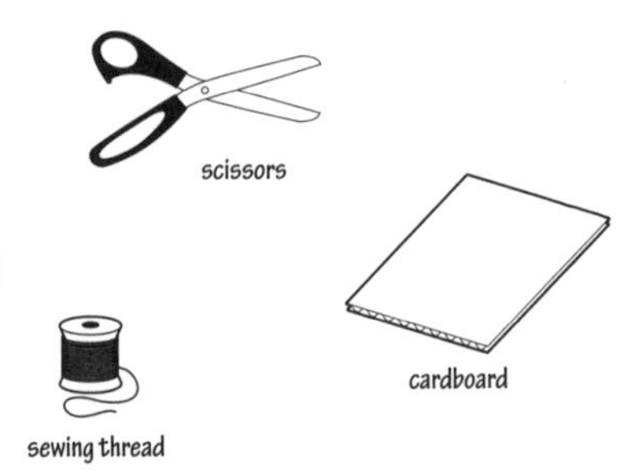

What to Do

Cut out the small shape shown in **Figure 1** from the piece of cardboard. Follow the dimensions shown. Next, try to balance the head of the figure on your fingertip, as shown in **Figure 2**. It's almost impossible to keep it upright without its tipping over.

Next, cut out the figure shown in **Figure 3**. The only difference is the legs are much longer. Try to balance this larger figure on your hand. It's easy now, because the center of gravity is below your finger. See **Figure 4**. You should be able to walk around the room and the figure will not fall.

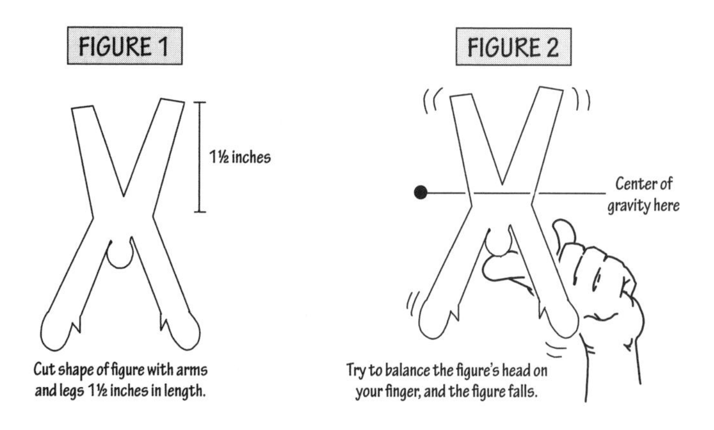

Cut shape of figure with arms and legs 1½ inches in length.

Try to balance the figure's head on your finger, and the figure falls.

Going Further

To demonstrate how acrobats keep their balance, cut a small slit in the head of the figure. See **Figure 5**.

Then, tie a length of thread from a chair to a lower object, such as another chair or table, and set the figure on the thread. The figure should rest on the thread in its slit and, with a slight push, slide across without falling. See **Figure 6**.

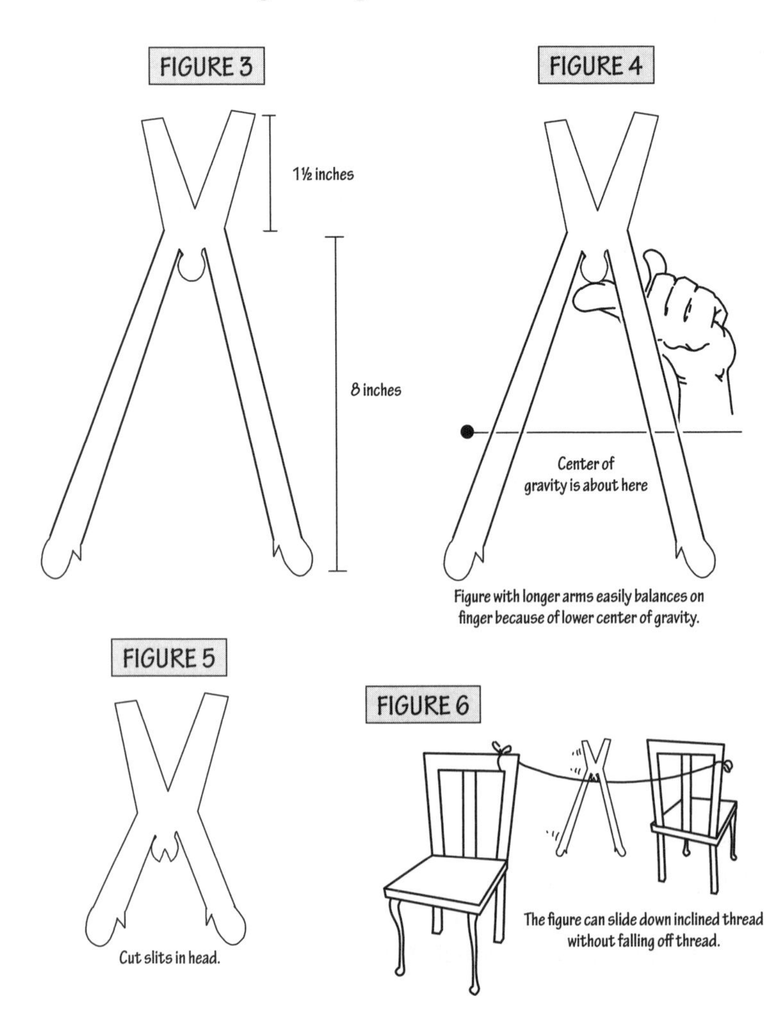

Figure with longer arms easily balances on finger because of lower center of gravity.

Cut slits in head.

The figure can slide down inclined thread without falling off thread.

Sneaky Balancer II

This sneaky balancer can rest horizontally on the tip of a paper clip and will surely astonish onlookers.

What's Needed

- Scissors
- Cardboard
- Paper clip

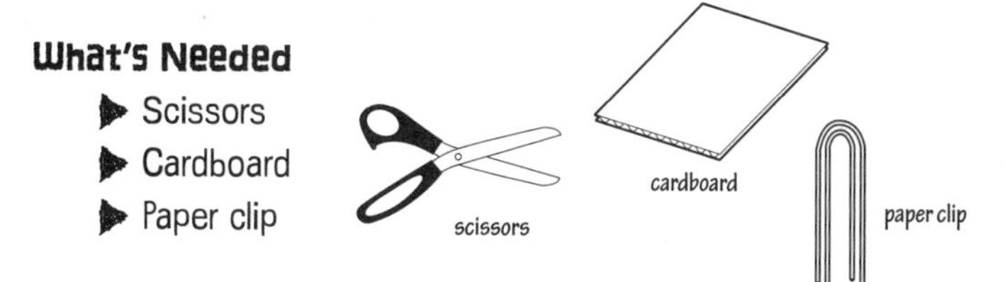

What to Do

Cut out the figure shown in **Figure 1** from the piece of cardboard. Be sure to include the spiked hair, with a long center spike. Try to adhere to the dimensions shown but, if desired, you can produce a larger or smaller figure as long as you keep the arm and body lengths in proportion.

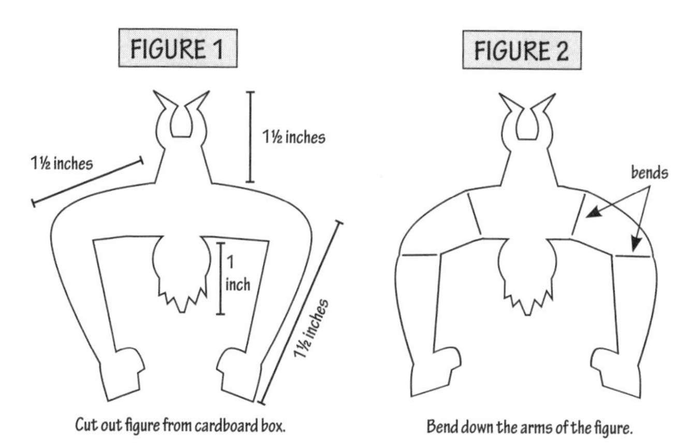

Cut out figure from cardboard box.

Bend down the arms of the figure.

Bend the figure's arms down at the shoulder and elbows. See **Figure 2**.

Next, bend a paper clip, as shown in **Figure 3**, so one end stands up vertically.

Last, place the figure on the paper clip with the spiked hair resting on the tip. If necessary, bend the arms down so it won't fall. The figure should balance on the tip of the paper clip. You should be able to carefully push its legs to the right or left and it will stay aloft. See **Figure 4**.

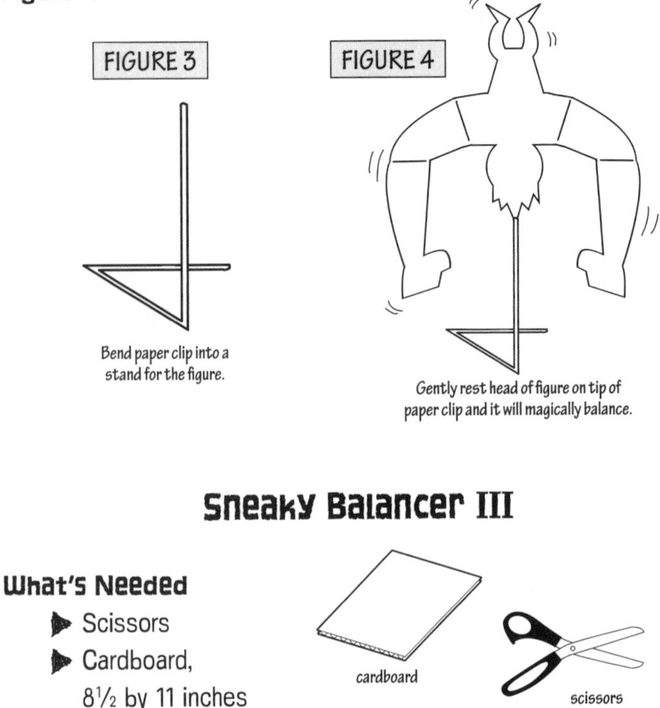

Bend paper clip into a stand for the figure.

Gently rest head of figure on tip of paper clip and it will magically balance.

Sneaky Balancer III

What's Needed

- Scissors
- Cardboard, $8\frac{1}{2}$ by 11 inches

What to Do

Cut out the shape shown in **Figure 1** from the cardboard. Be careful to follow the dimensions shown.

You should be able to easily balance the figure on the tip of your finger, elbow, or nose, because its center of gravity is at the large circular area. See **Figure 2**.

You can create similar figures and, using paper clips or coins secured with tape, add weight to an area near the bottom section of the figure so it balances effortlessly.

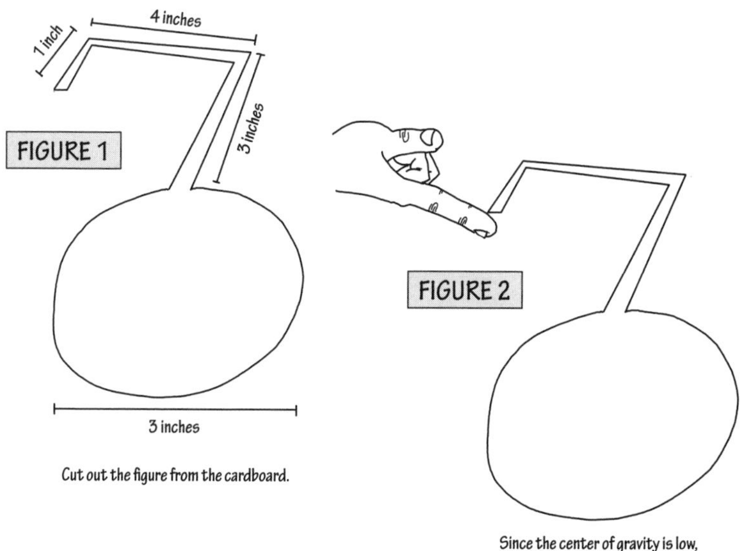

Cut out the figure from the cardboard.

Since the center of gravity is low, the figure will balance easily.

Part II

Sneaky Science Projects

Science is sometimes difficult to understand but you can demonstrate its principles with common household items. Paper and cardboard from product packaging, paper clips, aluminum foil, and paper cups can be quickly transformed into practical science projects.

If you're curious about the sneaky adaptation possibilities of more complex household devices, you're in the right place. People frequently throw away damaged gadgets and toys without realizing they can serve unintended purposes. You'll learn sneaky sources for wire and how to connect things. You'll see how to make a Frisbee disc with ordinary paper, clever center-of-gravity balancing designs, a handheld and a palm-sized cardboard boomerang, a sneaky periscope from a cookie box, and much more.

All the projects have been tested and can be made safely in no time. If you want to practice recycling and learn high-tech resourcefulness, the following projects will provide plenty of chances for fun product reuse applications.

Sneaky Periscope

One of the most useful devices you can build is a periscope, which allows you to take sneak peeks around corners or over fences without being seen. A periscope uses two mirrors positioned so light is reflected from the top mirror down to the lower mirror.

This project illustrates how to construct your own sneaky periscope in no time using an ordinary food carton and a couple of mirrors or watch batteries.

What's Needed

- Two small mirrors or watch batteries
- Long cardboard food box, typically 8 inches long and 1 inch wide
- Transparent tape
- Scissors

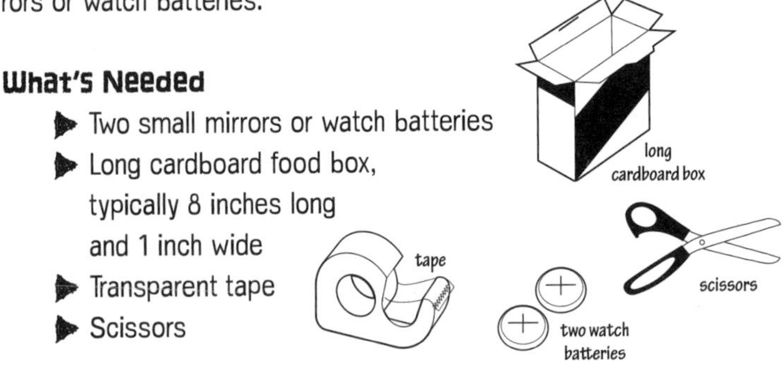

What to Do

First, unfold the food container box and lay it flat as shown in **Figure 1.**

Next, cut two 1 -inch square holes in the box as shown in **Figure 2.** Cut one hole close to the top end of section 1 and cut the other hole near the bottom edge of section 3.

At the opposite ends of the holes in sections 1 and 3, fold over the flaps to form triangles and secure them to the box with tape. See **Figure 3.** These flaps will act as mirror mounts.

Position the mirrors (or watch batteries) on the triangular flaps so they are leaning toward the center of the cardboard and secure them with tape as shown in **Figure 4**.

Last, fold the box so it's back to its original form. Secure the seams and ends with tape. See **Figure 5**.

When you hold the periscope upright, as shown in **Figure 6**, you will be able to see over tall obstructions without being seen. When you hold the periscope sideways, you will be able to see around corners.

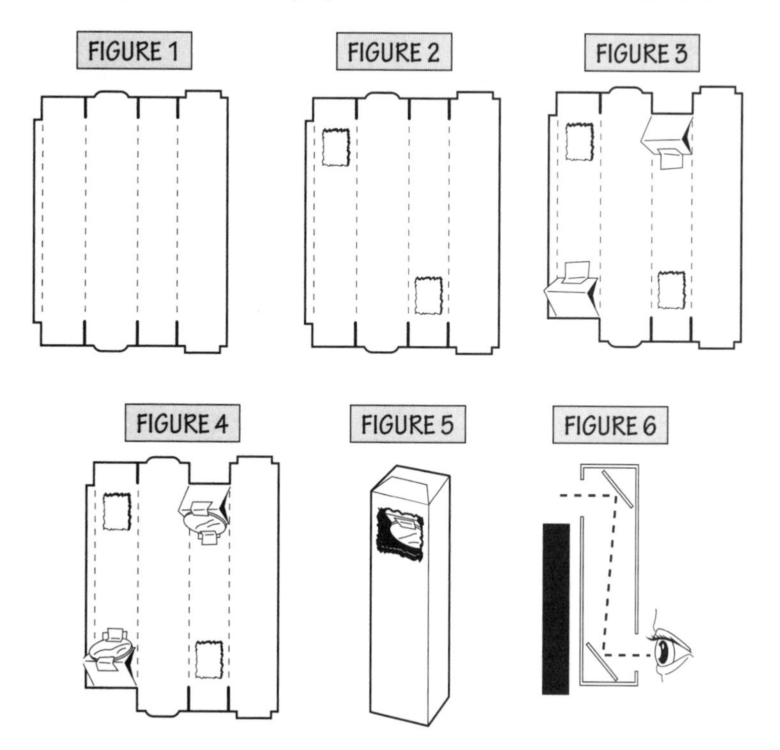

Sneaky Origami

Paper folding is fun but you can enhance your enjoyment by making the following sneaky origami designs that include motion action using everyday things.

Sneaky Head-Bobbing Bird

What's Needed

- Scissors
- Paper
- Pencil

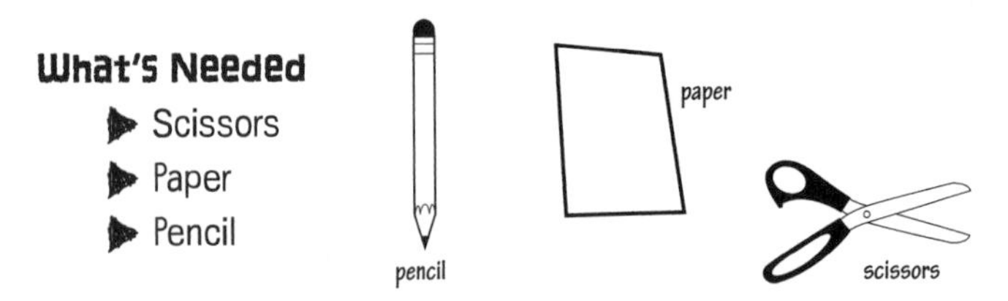

What to Do

Cut the paper into a square and fold/unfold both the diagonals, as shown in **Figure 1**. Fold over the top left and right corners to the center. See **Figure 2**.

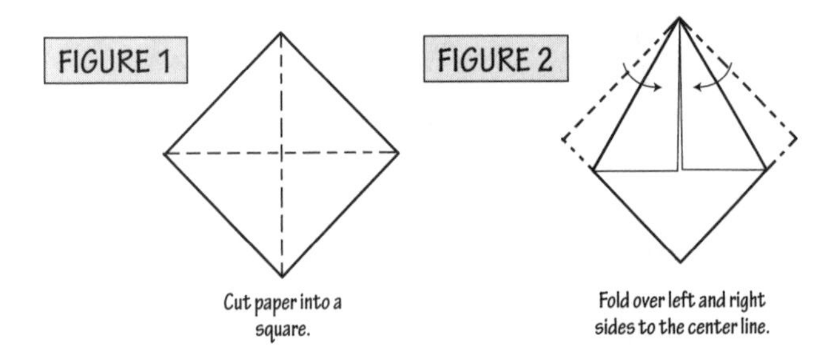

Cut paper into a square.

Fold over left and right sides to the center line.

Then, fold over the lower left and right corners toward the center as shown in **Figure 3**. Fold up the bottom point to the center line to form a tail and fold the top corner toward the back of the figure to make the head, as shown in **Figures 4** and **5**.

Next, fold the figure in half vertically along the center toward the tail. This will bend the tail and head outward as shown in **Figure 6**. Draw eyes and a beak on the figure as desired.

Last, with the sneaky bird standing upright, push down on the center of the tail. The head should move downward. See **Figure 7**.

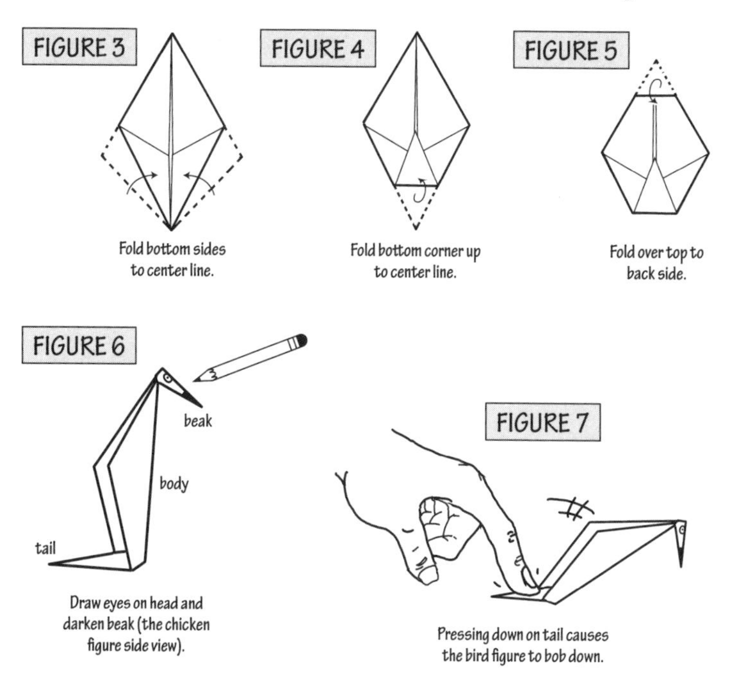

Fold bottom sides to center line.

Fold bottom corner up to center line.

Fold over top to back side.

Draw eyes on head and darken beak (the chicken figure side view).

Pressing down on tail causes the bird figure to bob down.

Sneaky Mouth Flapper

What's Needed

- Scissors
- Paper
- Pencil

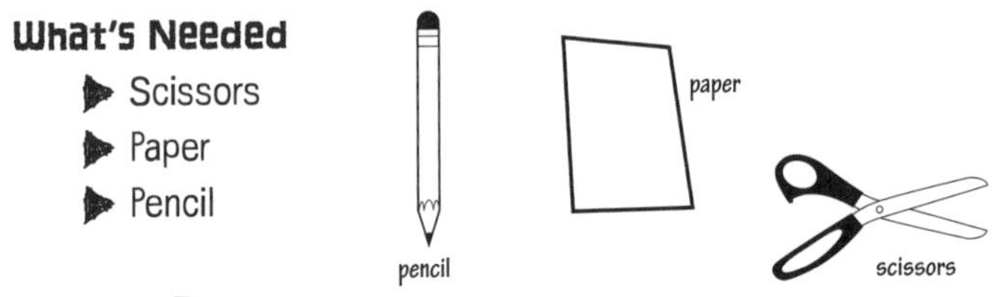

What to Do

Cut the paper into a square, as shown in **Figure 1**. Next, fold and unfold the square on both the diagonals. Fold over the lower left and right corners to the center. See **Figure 2.**

Then, fold over the upper left and right corners toward the center, as shown in **Figure 3.** Fold up the bottom half of the figure along the center crease. See **Figure 4.** Fold down the top front corner to the bottom of the figure, as shown in **Figure 5.**

Fold the top back tip down and to the left along the fold line shown in **Figure 5.** Then, fold the bottom front tip up and to the left along its indicated diagonal fold line, until it resembles the shape in **Figure 6.** Unfold the left-pointing tips back to their positions shown in **Figure 5.** See **Figure 7.**

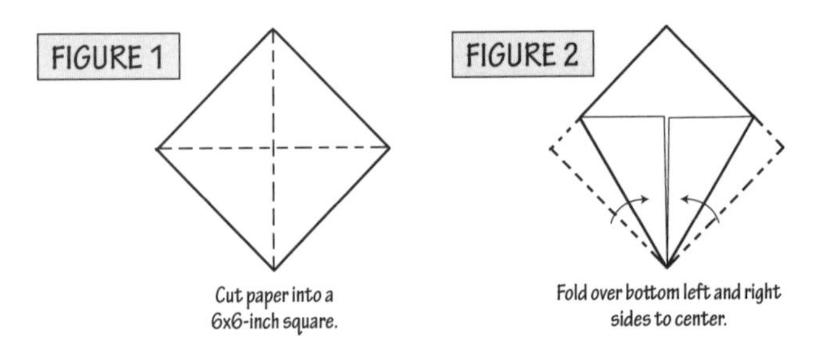

Cut paper into a 6x6-inch square.

Fold over bottom left and right sides to center.

Next, fold the top corner down and to the right in the opposite direction of how you folded it to make **Figure 6**. Similarly, fold the bottom corner up but to the right, until the shape appears like the one in **Figure 8**.

Fold the bottom right corner to the center—it will fold the figure in half. See **Figure 9**. While you are folding it, shape the top right-pointing corners into a mouth shape by pushing the beak with your hand. See **Figure 10**. If necessary, fold and unfold the figure until this section resembles a mouth.

Last, draw eyes on both sides of the top portion of the figure. Now you can pull on the two bottom corners and the mouth will flap open and close, as shown in **Figure 11**. If not, unfold the beak and refold it while adjusting it with your hand until the mouth moves properly.

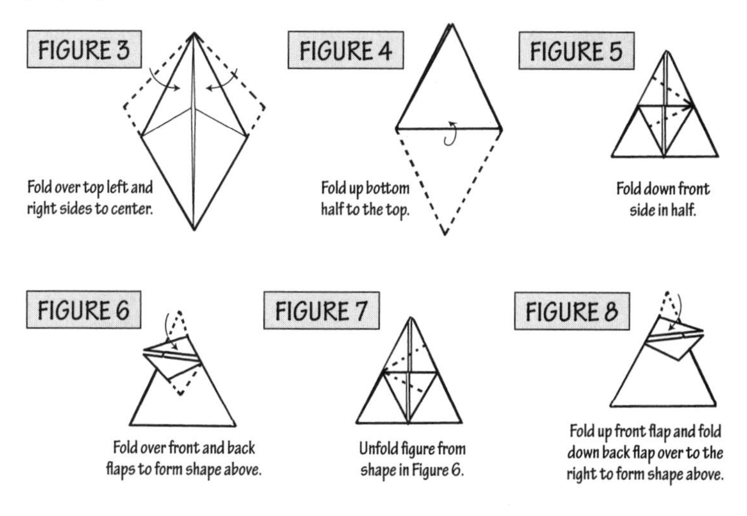

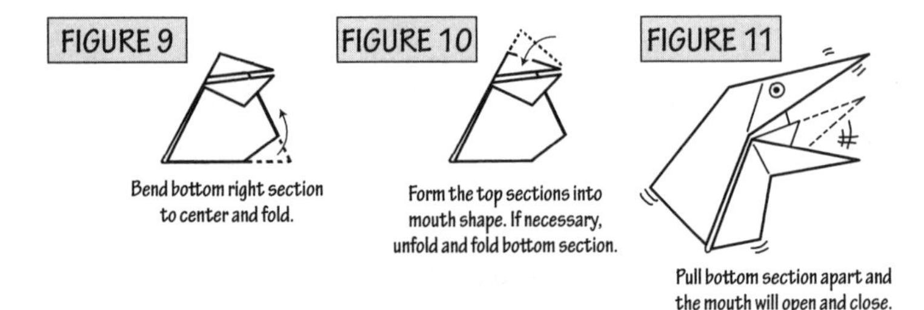

Sneaky Origami Animator

You can add motion to your origami designs, and other craft creations, by making a Sneaky Origami Animator with everyday objects.

What's Needed

- Two large paper clips
- Electrical tape
- Five by three-inch piece of cardboard
- Needle-nose pliers

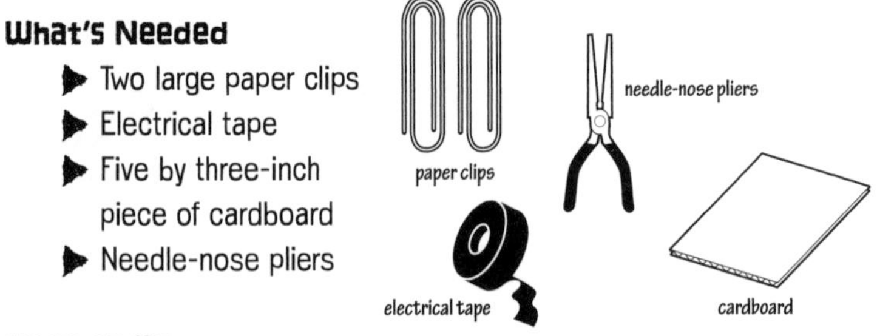

What to Do

This project illustrates how to make a cam-crank toy to add locomotion to your still figure designs. You can produce variations on this design by using larger pieces of cardboard and stiff wire, but it's recommended to make a simple version first. Later, you can alter the size of the parts to produce your desired results.

First, bend one paper clip into the shape shown in **Figure 1**. It will act as a mount for your origami figure.

Next, bend the second paper clip into the shape shown in **Figure 2.** It will act as a cranking cam that will move the first paper clip up and down. Wrap electrical tape around both paper clips.

Poke holes in the cardboard at 1-inch, 2½-inch, and 4-inch intervals, as shown in **Figure 3.** Then, stand the card along its long side and fold it into a U shape.

Push the first paper clip into the center hole. Use pliers to bend the top of the clip into a C shape so it will not fall through the hole. See **Figures 4** and **5.**

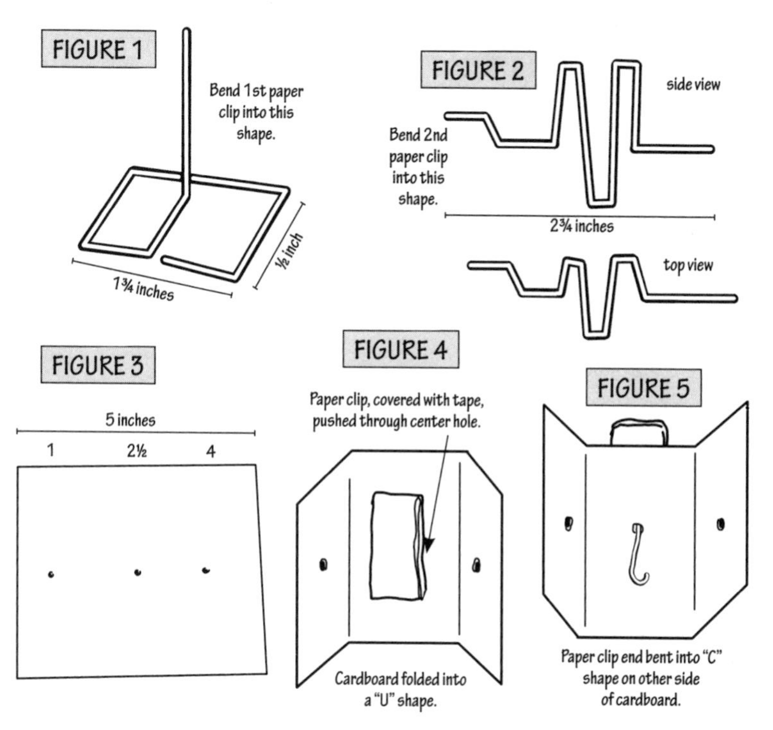

Next, push the second paper clip into the side holes of the cardboard so it rests underneath the first paper clip, as shown in **Figure 6**. Apply tape to the bottom of the cardboard to keep its shape.

Last, turn the paper clip crank on the side of the Sneaky Origami Animator and the top paper clip will move up and down. See **Figure 7**. Since the first paper clip has an irregular shape, it acts as a cam mechanism and causes erratic movement on the other paper clip resting on it.

You can attach small paper figures to the top paper clip with tape. Experiment with an assortment of shapes for your paper clip cam (e.g., oval or triangular shapes) to produce a variety of motion effects. See the moving-arm figure in **Figure 8**.

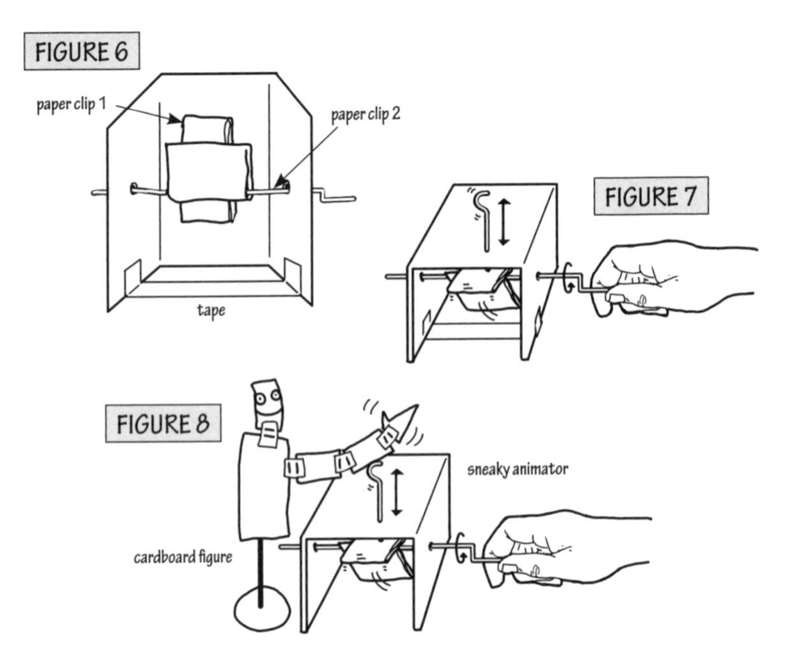

Bernoulli Principle Tricks

Sneaky Demonstrations of Air Pressure and Wing Lift

Have you ever wondered how airplanes and helicopters are able to fly? If you have, and want to demonstrate this principle, all you need are such ordinary items as straws, postcards, and strips of paper.

Air Pressure Demonstration I

An ordinary straw can be used to demonstrate that air pressure is all around us (15 pounds per square inch, to be exact). You can demonstrate this easily enough with everyday items.

What's Needed

- Straw
- Glass filled with water

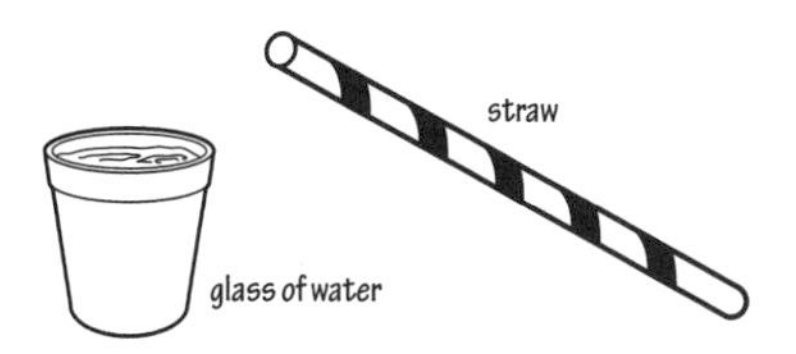

What to Do

Insert a straw into the glass of water, as shown in **Figure 1.** Next, place a finger over the top of the straw and lift it out of the water. See **Figure 2.**

You'll see that the water stays in the straw and doesn't flow out because air pressure from the bottom is keeping it in, as shown in **Figure 3.** When you lift your finger from the top of the straw, air pressure flows from the top and pushes against the water, forcing it out.

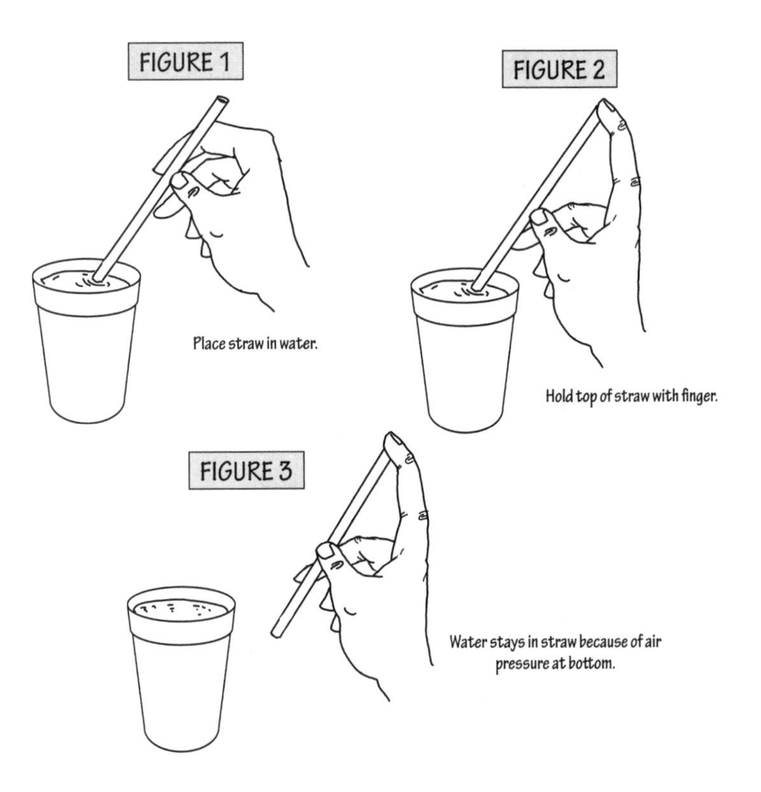

Air Pressure Demonstration II

You can demonstrate the power of air pressure in a more dramatic way with the following project, again using everyday items.

What's Needed

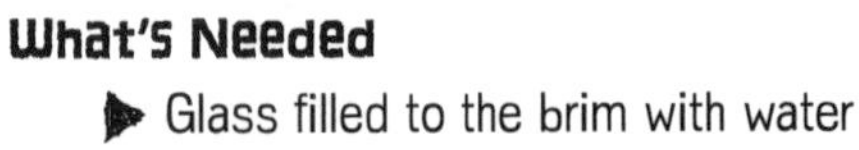

- Glass filled to the brim with water
- Plastic-coated postcard

What to Do

Working over a sink, hold up the glass of water. Place a postcard over the mouth of the glass and turn the glass upside down, holding the postcard in place with your finger under it, as shown in **Figure 1**.

Carefully remove your finger from the postcard and you should see that the postcard will not fall. With no air in the glass to push against the postcard, the air outside presses against the postcard, keeping it in place, even with the weight of the water upon it. See **Figure 2**.

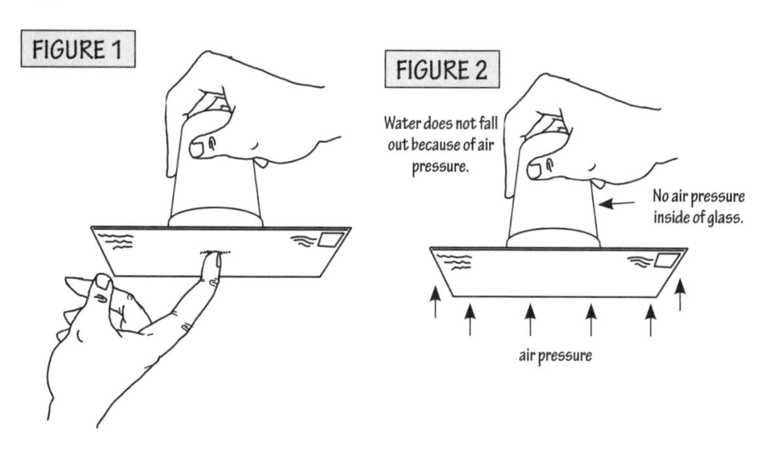

Air Pressure Demonstration III

What's Needed

- Paper (preferably a paper towel or napkin)
- Scissors

What to Do

Cut a paper strip ½ inch wide by 4 inches in length as shown in **Figure 1**. Hold the paper strip up to your face above your mouth and blow. The paper naturally moves upward. Now hold the paper strip just below your lips and blow above the strip. As shown in **Figure 2**, the paper will also rise and move upward!

This occurs because of Bernoulli's principle, which states that fast-moving air has less pressure than nonmoving air. The air under the strip has more pressure than the air above it and pushes the strip upward.

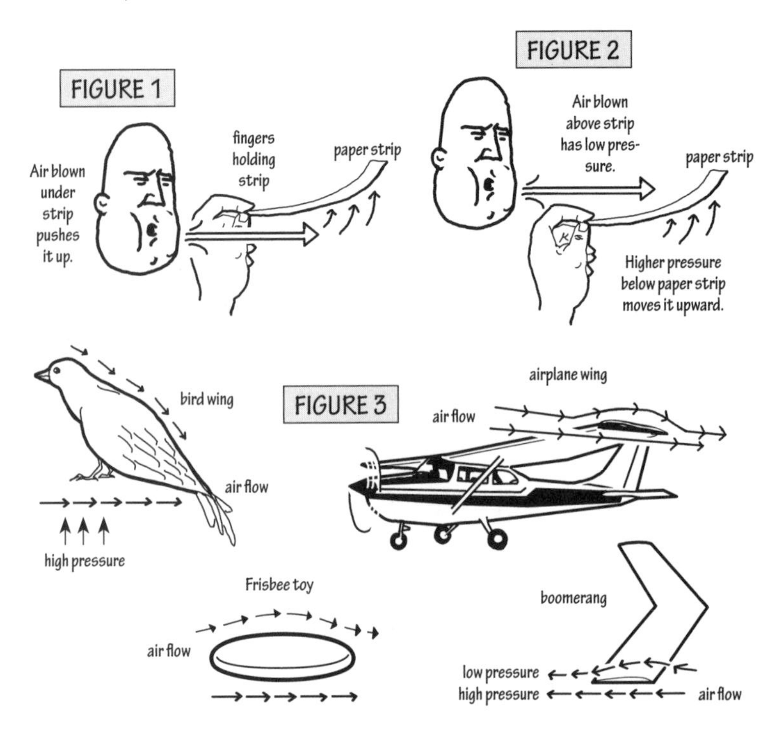

Figure 3 illustrates a side view of a bird's wing, an airplane wing, a Frisbee flying disk, and a boomerang. Notice the top of the wing curves upward and has a longer surface as compared to the bottom. When the airplane moves forward, air moves above and below the wing. The air moving along the curved top must travel farther and faster than the air moving past the flat bottom surface. The faster-moving air has less pressure than the air at the bottom and this provides lift.

Baseball pitchers can take advantage of Bernoulli's principle by releasing the ball with a forward spin. The ball produces a lower pressure below it, causing it to dip when it reaches the plate. Hence, a curveball. See **Figure 4**.

Sailboats apply Bernoulli's principle to use the wind, regardless of its direction, to propel the boat in any desired direction. **Figure 5** shows how altering the shape of the sail into a curve produces an effect similar to that of an airplane wing. The wind moves at a faster rate over the curved side, with a lower pressure, and the higher pressure on the other side of the sail pushes the boat laterally. A centerboard, attached to the boat hull, prevents the boat from moving sideways while allowing it to use the wind thrust to move forward. See **Figure 6**.

Automobile bodies are similar to an airplane wing because they are flat on the bottom and curved on top. They can lose stability at high speeds since they tend to achieve lift from the higher air pressure below, as shown in **Figure 7**. To reduce the Bernoulli effect, automakers have incorporated improvements in vehicle design, such as lowering the body height, adding special front bumper and fender contours, and installing rear spoilers. See **Figure 8**.

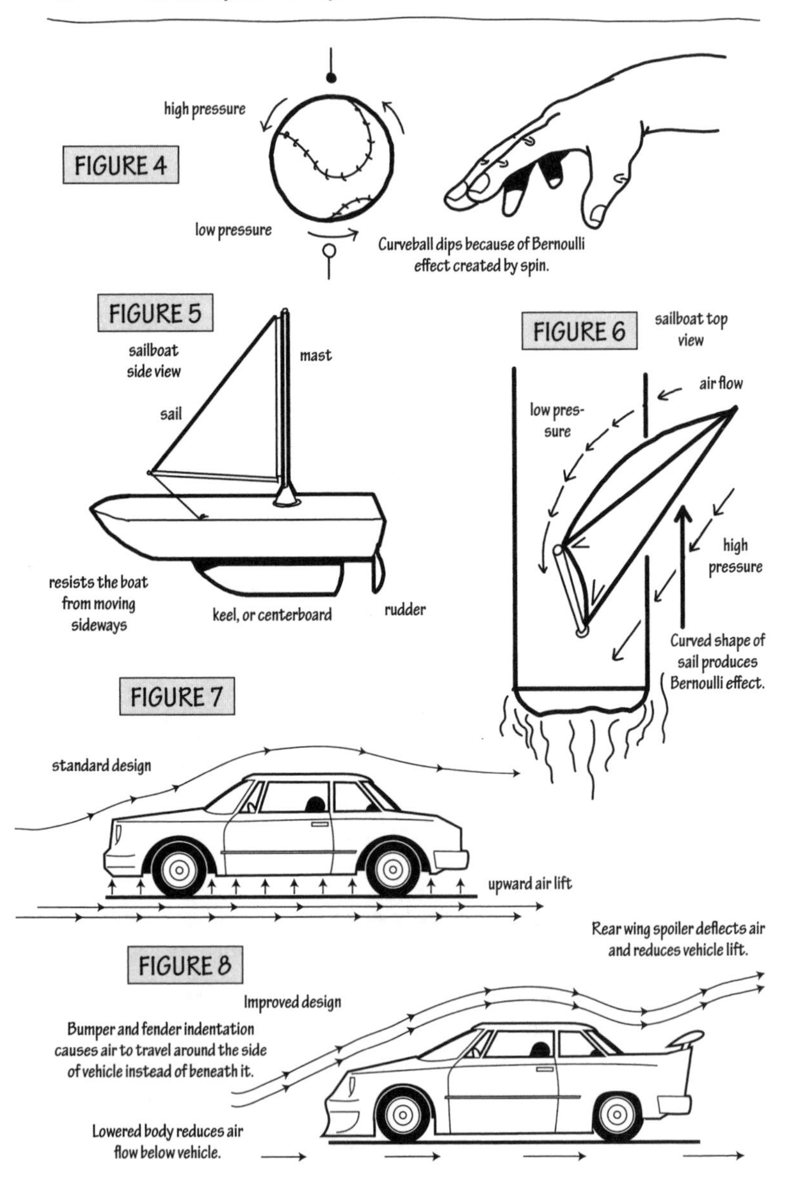
FIGURE 4
high pressure
low pressure
Curveball dips because of Bernoulli effect created by spin.
FIGURE 5
sailboat side view
mast
sail
resists the boat from moving sideways
keel, or centerboard
rudder
FIGURE 6
sailboat top view
air flow
low pressure
high pressure
Curved shape of sail produces Bernoulli effect.
FIGURE 7
standard design
upward air lift
FIGURE 8
Rear wing spoiler deflects air and reduces vehicle lift.
Improved design
Bumper and fender indentation causes air to travel around the side of vehicle instead of beneath it.
Lowered body reduces air flow below vehicle.

Air Pressure Demonstration IV

What's Needed

- Scissors
- Paper (preferably a paper towel or napkin)
- Two empty soda cans
- Magazine

What to Do

Cut two paper strips ½ inch wide by 4 inches in length and hold them about 2 inches apart, as shown in **Figure 1**. Blow air between the paper strips and watch what occurs. You would expect the strips to blow apart but they actually come together, as shown in **Figure 2**.

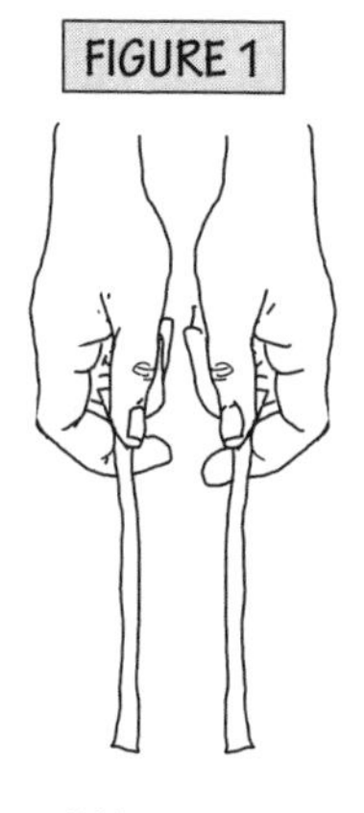

Hold paper strips 2 inches apart.

FIGURE 2

Air blown between paper srips moves faster with less pressure, causing them to move together.

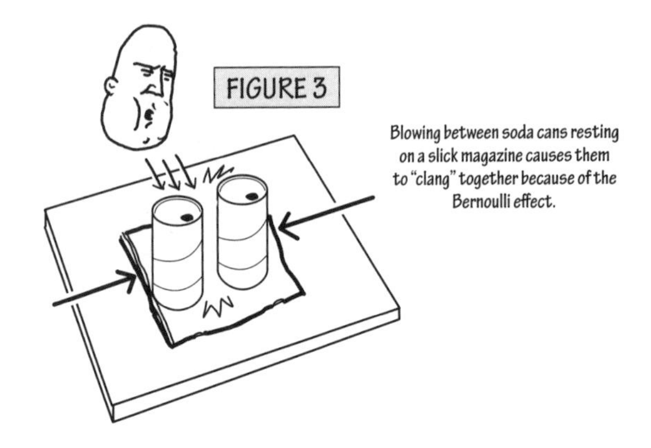

Blowing between soda cans resting on a slick magazine causes them to "clang" together because of the Bernoulli effect.

Bernoulli's principle is working here because the faster-moving air blown between the paper strips has less pressure than the air on the other side of the paper. This higher pressure pushes the strips toward each other.

Now, place the two empty soda cans an inch apart upon the slick surface of a magazine. When you blow between the cans, they will move toward each other, producing a clanging sound. See **Figure 3.**

Air Pressure Demonstration V

Here's another sneaky, easy-to-perform demonstration of air pressure's causing an unexpected result.

What's Needed

- Scissors
- Piece of paper

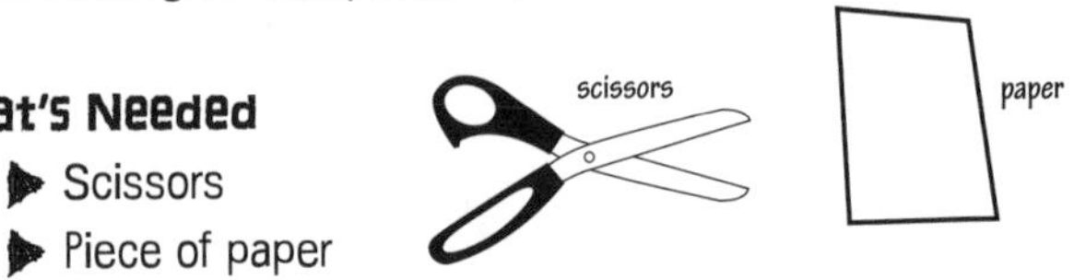

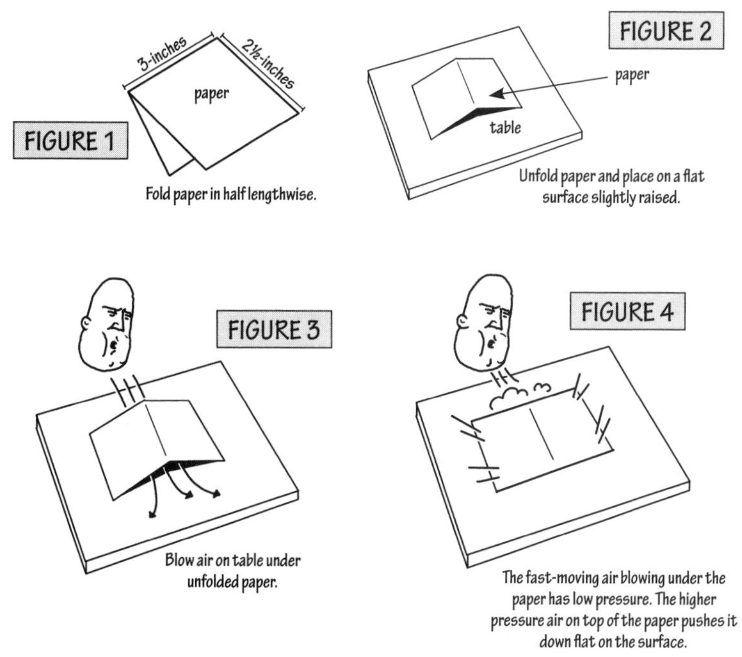

What to Do

Cut the piece of paper into a 5 by 3-inch shape. Fold the paper in half lengthwise, as shown in **Figure 1**.

Next, unfold the paper and place it on a flat surface so that it has a slight rise near its center crease. See **Figure 2**.

Then, as shown in **Figure 3**, bring your face close to the surface of the table and blow underneath the paper.

You would expect the paper to rise but it actually flattens downward. The higher air pressure on top of the paper, compared to the fast-moving air beneath it, pushes the paper flat on the table, as shown in **Figure 4**.

Air Pressure Demonstration VI

You can use Bernoulli's principle to perform a neat magic trick by making a ball rise from a cup and jump into another one without touching it.

What's Needed

- Ping-Pong ball
- Two small cups

What to Do

This project requires small cups that are slightly smaller in diameter than the Ping-Pong ball. Since the Ping-Pong ball can barely fit in the cup, rapidly moving air above the ball will not affect the air pressure beneath it.

Put the ball into one of the cups and place it about 3 inches away from the second cup, as shown in **Figure 1.** Blow as hard as you can above the first cup and the ball should start to rise. See **Figure 2.** The force of your breath will push the raised Ping-Pong ball over to the empty cup, where it will drop inside, as shown in **Figure 3.** With a little practice, you can make this sneaky trick work every time.

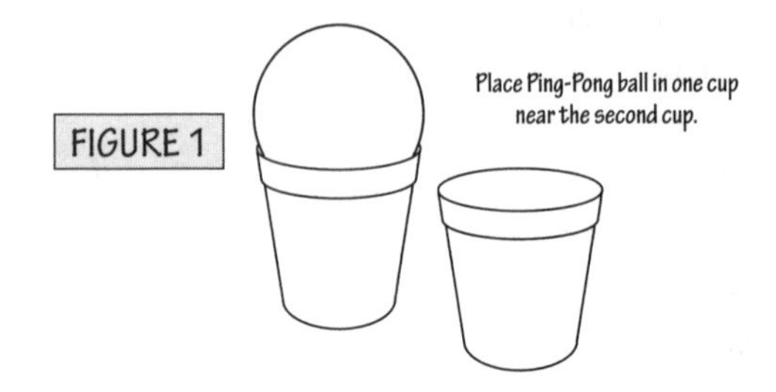

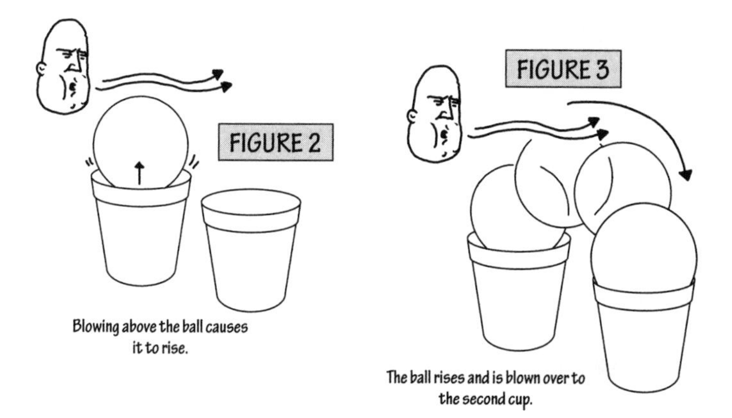

Blowing above the ball causes it to rise.

The ball rises and is blown over to the second cup.

Sneaky Frisbee from Paper

Sneaky Flying Disk

You've seen how Bernoulli's principle works. Now it's time to put it to use and make a sneaky flyer, similar to flying disk toys, using paper and tape.

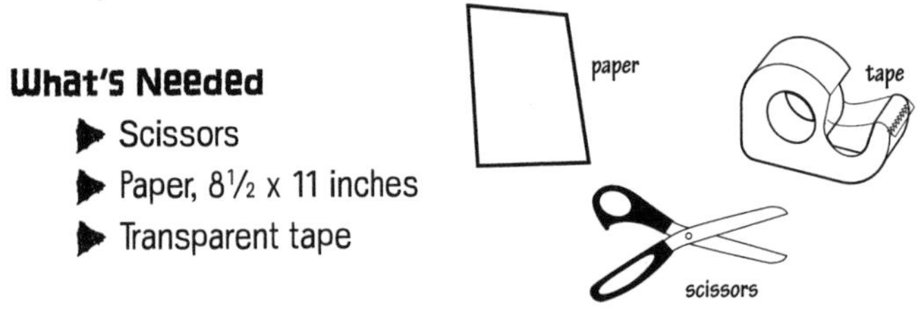

What's Needed

- Scissors
- Paper, 8½ x 11 inches
- Transparent tape

What to Do

Cut eight 2-inch square pieces of paper as shown in **Figure 1**. Fold the top right corner of one square down to the lower left corner. See **Figure 2**. Then, fold the top left corner down to the bottom, as shown in **Figure 3**.

Repeat these two folds with the remaining seven squares. See **Figure 4**.

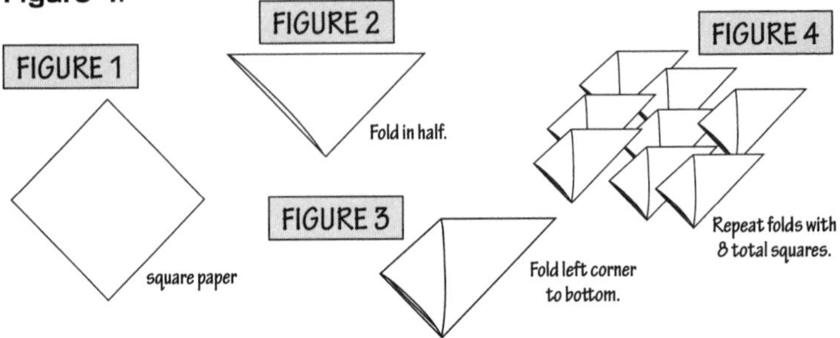

Insert one paper figure into the left pocket of another, as shown in **Figure 5**. Repeat inserting the figures into one another until they form an eight-sided doughnut shape; see **Figure 6**. Apply tape as needed to keep the origami flyer together and turn over, as shown in **Figure 7**.

Next, bend up the outer edge of the sneaky flyer to form a lip, as shown in **Figure 8**. This outer lip will cause the air to take a longer path over it, producing a Bernoulli effect.

Turn the device so the lip is bent downward. Throw the Sneaky Flying Disk with a quick snap of your wrist and it should stay aloft for a great distance.

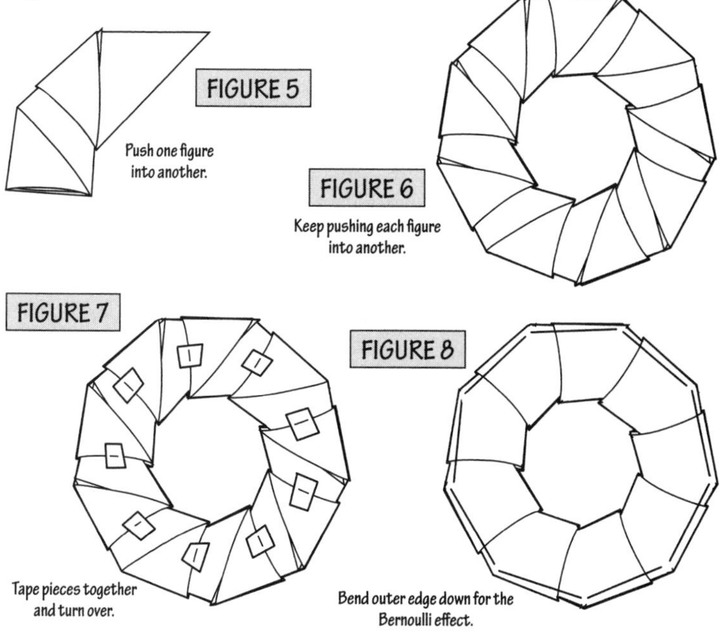

Sneaky Boomerang

Want a sneaky way to play catch alone? You just need a piece of cardboard and foam rubber to make a working boomerang that will actually fly up to 30 feet away and return.

What's Needed

- Scissors
- Cardboard from a food box
- Foam rubber, from an old pillow
- Transparent tape

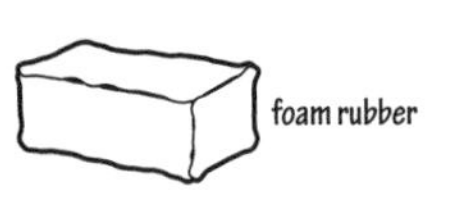

What to Do

Cut the cardboard into the boomerang shape shown in **Figure 1**. Each wing of the boomerang should be 9 inches long by 2 inches wide.

Then, cut two foam pieces into 6 by 2-inch oval shapes with one side rising into a curve. The rising shape should resemble the side view of an airplane wing. See **Figure 2**. Place the oval foam pieces on the leading edges of the boomerang and secure them with tape.

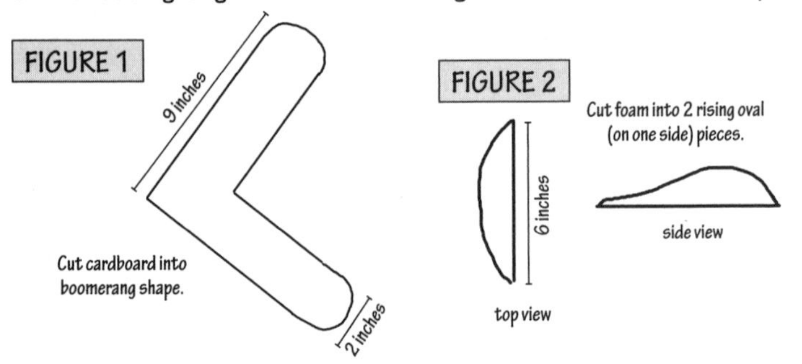

Note: Look carefully at the placement of the ovals on the boomerang wings in **Figure 3** before taping them. The foam creates a curved shape on the boomerang wing, which will cause air to move faster across its top than across the bottom surface. This will produce lift for the boomerang.

Hold the boomerang as if you were going to throw a baseball and throw it straight overhead (not to the side). See **Figure 4**. The Sneaky Boomerang should fly straight and return to the left. Experiment with different angles of throw to obtain a desired return pattern.

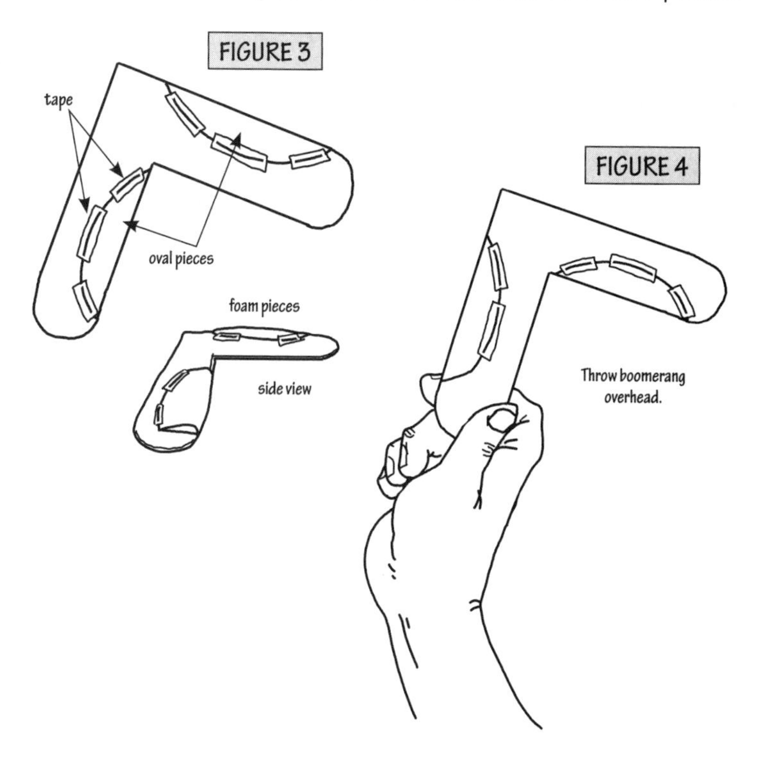

Sneaky Mini-Boomerang

You can use postcards, business cards, or cardboard food boxes to make a miniature, palm-size boomerang that actually flies and returns to you, for indoor fun.

What's Needed

- Scissors
- Cardboard from food boxes or postcards

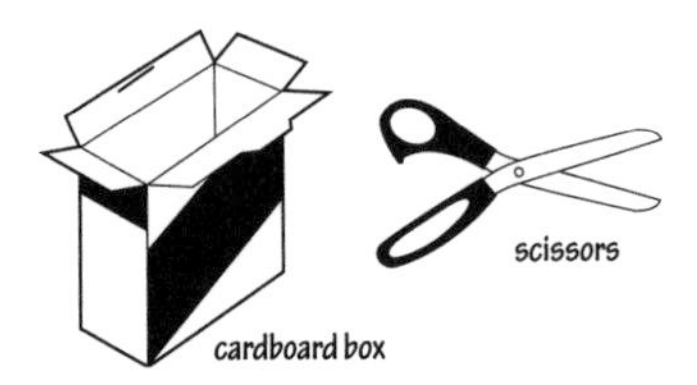

What to Do

Cut out the boomerang shapes shown in **Figure 1**. The boomerang wings can be any length between 2 to 4 inches. For optimal flight height and return performance, cut each wing of the boomerang 2½ inches long and ½ inch wide.

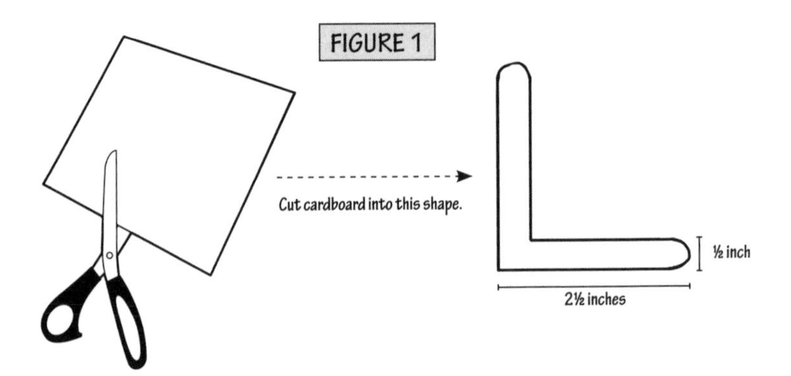

Set the Sneaky Mini-Boomerang on the palm of your raised hand with one wing hanging off. Tilt your hand slightly upward. With your other hand's thumb and middle finger ½ inch away, snap the outer boomerang wing. You'll discover (after a few attempts) that it will fly forward and return to you. See **Figure 2.**

Note: You must snap your finger with a strong snapping action to make the boomerang fly away and return properly, as shown in **Figure 3.**

Experiment with different hand positions and angles to control the boomerang's flight pattern.

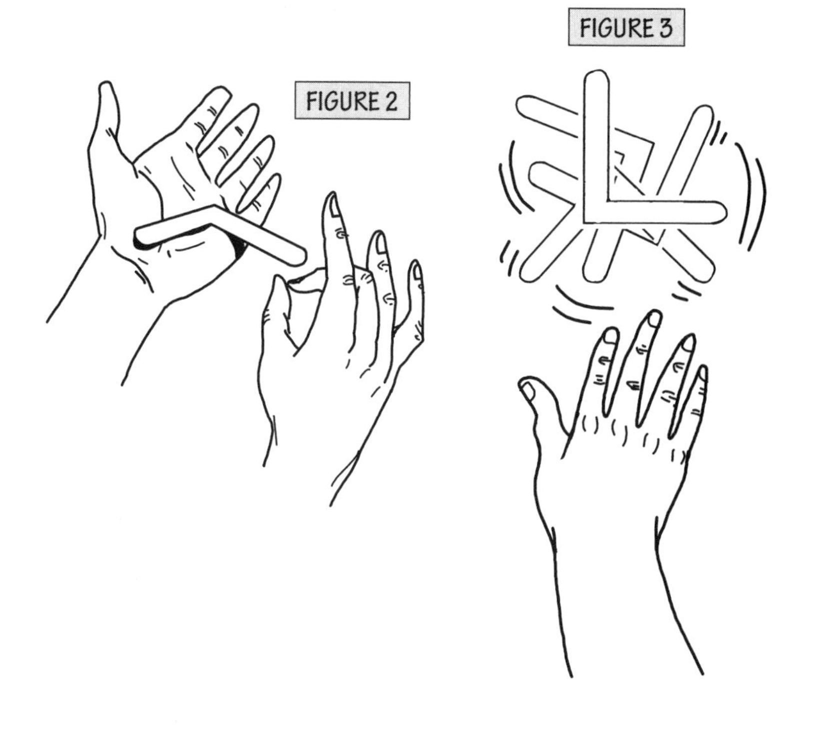

Sneaky Gliders

You don't have to spend money on a balsa wood kit to make a simple working glider. A working glider, made from discarded cardboard or Styrofoam material, can produce plenty of sneaky flyers for safe fun.

What's Needed

- Scissors
- Flat corrugated cardboard or Styrofoam
- Transparent tape

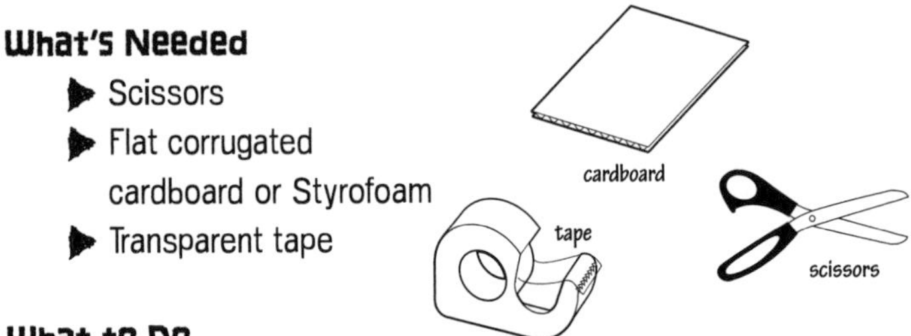

What to Do

The sneaky glider body, or fuselage, can be cut out from the pattern shown in **Figure 1**. The plane will require at least one wing near the center for stability. A smaller wing near the rear rudder can also be added. Simply insert the wing(s) into the body slits and use tape to secure them properly as shown in **Figure 2**.

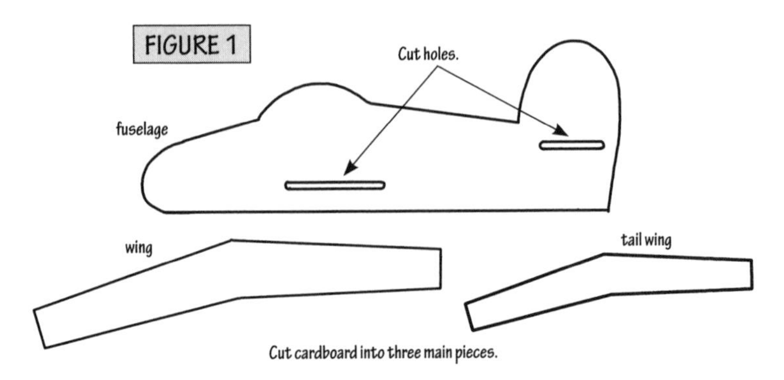

Cut cardboard into three main pieces.

Launch the Sneaky Glider with a snap of the wrist near your ear and it should fly up to 30 feet away. See **Figure 3**. Test the glider wing(s) shapes to achieve various flight paths as desired.

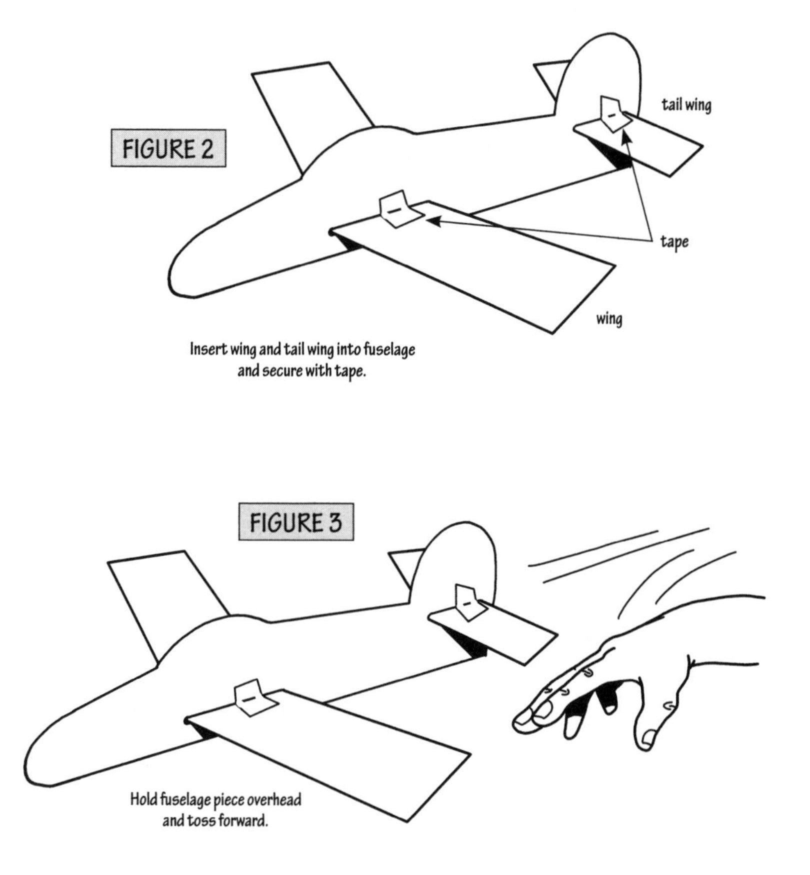

Insert wing and tail wing into fuselage and secure with tape.

Hold fuselage piece overhead and toss forward.

Sneaky Hoop Paper Flyer

Paper airplane designs are not hard to find. But if you want to stand out from the crowd, make this unique sneaky flyer using just a straw and paper.

What's Needed

- Scissors
- Sheet of paper
- Tape
- Straight drinking straw

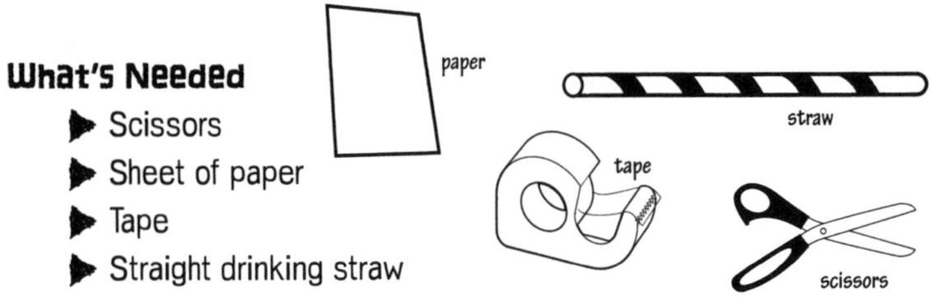

What to Do

First, cut two paper strips ½ inch wide by 4 inches long and then tape each strip into a loop, as shown in **Figure 1**. Next, tape a loop to each end of the straw. See **Figure 2**.

Now launch the sneaky straw flyer with your hand as if you were throwing a dart. It should fly up to 40 feet away.

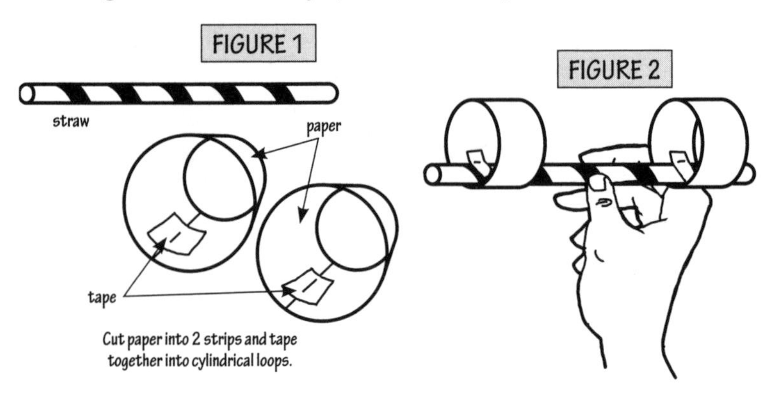

Cut paper into 2 strips and tape together into cylindrical loops.

Sneaky Soaring Cylinder

Paper airplanes don't have to have a standard-looking shape to glide long distances. Believe it or not, you can fold an ordinary piece of paper into the shape of a cup and amaze your friends with a Sneaky Soaring Cylinder.

What's Needed

- Sheet of paper

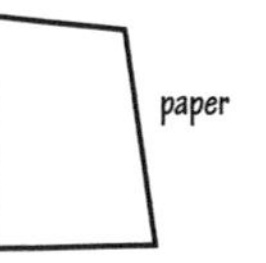

What to Do

First, fold the left side of the paper 2 inches to the right, as shown in **Figure 1**. Next, fold the paper from the left side one more inch, and crease it firmly. See **Figure 2**.

Roll the paper into a cylinder, then slide one end of the paper into the other end's folded-over area, as shown in **Figure 3**. Push the left side of the paper into the right side until about two inches' worth is securely in place. Then, roll over and firmly crease the edge of the folded paper into a lip to secure it. See **Figures 4** and **5**.

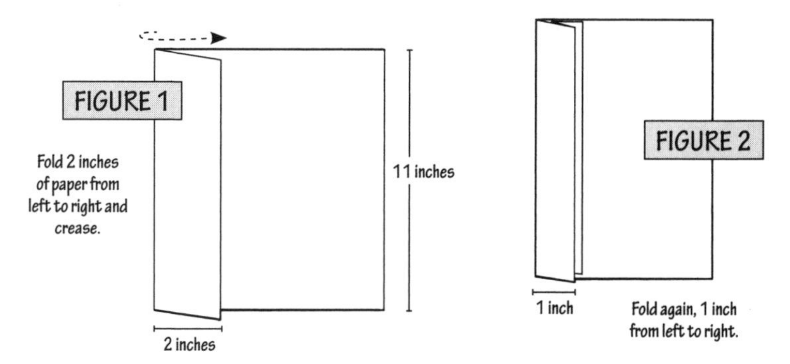

Next, bend over the top layer of paper into a fin shape, shown in **Figure 6**, so it stands vertically. This will act as a wind stabilizer to keep the cylinder in the air.

Last, toss the cylinder like a football, but don't add a spinning motion. See **Figure 7**. The Sneaky Soaring Cylinder should fly up to 40 feet away. Experiment with the shape of the stabilizer fin to achieve the desired various flight paths.

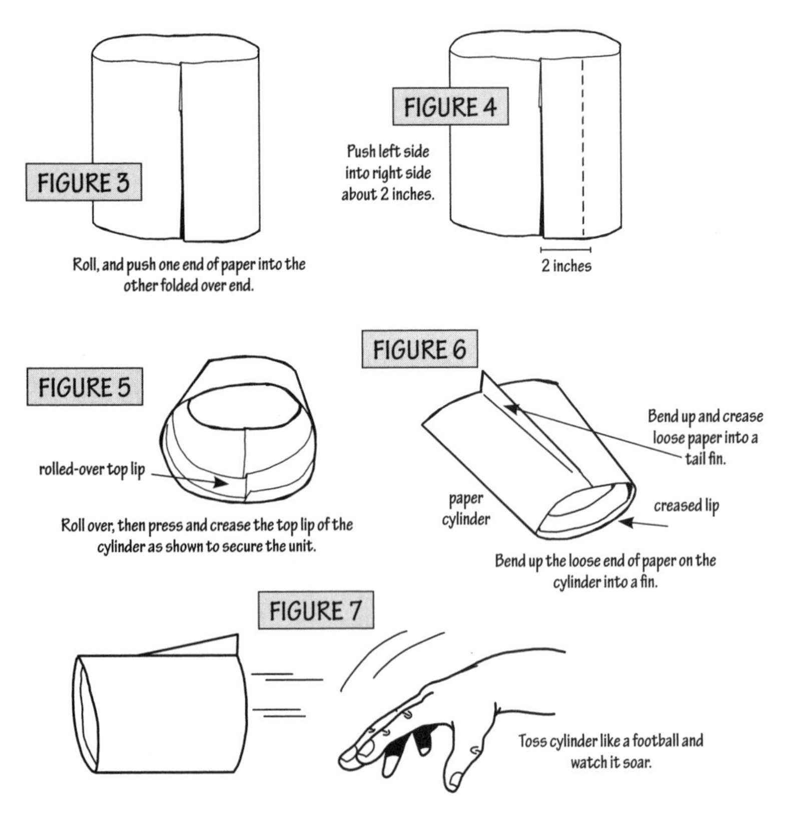

Sneaky Animated Cassette Tape Creations

Magnetism is used to record and play back tape recordings. Tiny iron particles on recording tape align themselves according to the signal placed there by the magnetic tape recording heads. On play back, the now-magnetized material in the tape moves across the tape head, which has coils of wire inside, and the signal is detected and amplified by the recorder to produce sound.

You can make some sneaky craft designs with strips of tape and animate them with a strong magnet.

What's Needed

- Cassette tape
- Strong magnet
- Cardboard
- Glue
- Scissors

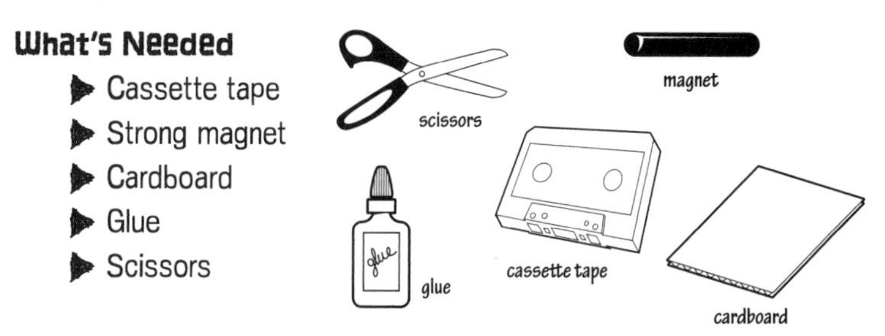

What to Do

First, draw a figure of your choosing. In **Figure 1**, an illustration of a man is shown.

FIGURE 1

Next, cut 2- to 3-inch strips of cassette tape and place glue on one end of the tape strips. Press the strips on the figure to create hair and a beard as shown in **Figure 2**. Let the sneaky design dry for 30 minutes.

FIGURE 2

Then, bring a magnet close to the drawing near the top of the figure's head. As shown in **Figure 3**, the "hair" made from the cassette tape will stand up. Try different drawings and tape arrangements to see what other animated illustrations you can create.

FIGURE 3

Make Wire and Batteries in a Pinch

In an emergency, you can obtain wire—or items that can be used as wire—from some very unlikely sources.

To test an item's conductivity (its ability to let electricity flow through it), use a flashlight bulb or an LED. LED is short for light-emitting diode; LEDs are used in most electronic devices and toys as function indicators because they draw very little electrical current, operate with very little heat, and have no filament to burn out.

Lay a small 3-volt watch battery on the item to be tested, as shown in **Figure 1**. If the bulb LED lights, the item can be used as wire for battery-powered projects.

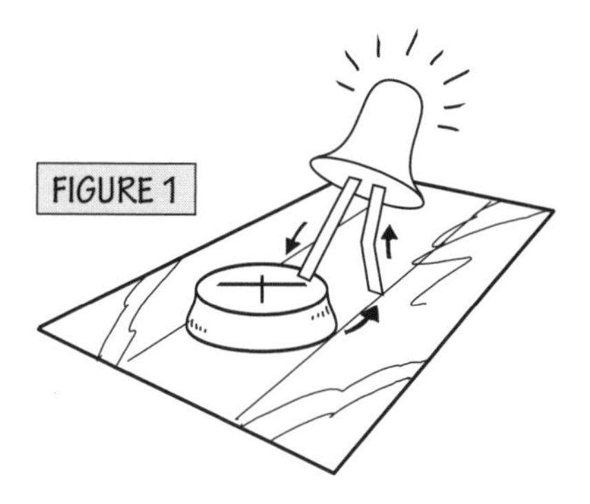

You can cut strips of aluminum material from food wrappers easily enough. With smaller items—such as aluminum obtained from a coffee-creamer-container lid—use the sneaky cutting pattern shown in **Figure 2**. Special care must be taken handling the fragile aluminum materials listed. In some instances, aluminum material will be covered by a wax or plastic coating that you may be able to remove.

Note: Wire from aluminum sources is only to be used for low-voltage battery-powered projects.

The resourceful use of items to make sneaky wire is not only intriguing, it's fun, too.

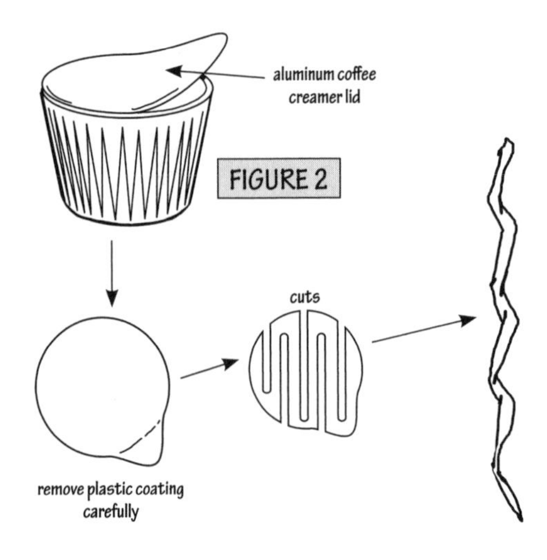

Sneaky Wire Sources

Ordinary wire can be used in many sneaky ways.

When wire is required for sneaky projects, whenever possible try to use everyday items that you might otherwise have thrown away. Recycling metal will help save our natural resources.

Getting Wired

In an emergency, you can obtain wire—or items that can be used as wire—from some very unlikely sources. **Figure 1** illustrates just a few of the possible items that you can use in case connecting wire is not available

Ready-to-use wire can be obtained from:

- Telephone cords
- TV/VCR cables
- Headphone wire
- Earphone wire
- Speaker wire
- Wire from inside toys, radios, and other electrical devices

Note: Some of the sources above will have one to six separate wires inside.

Wire for projects can also be made from:

- Take-out food container handles
- Twist-ties
- Paper clips

- Envelope clasps
- Ballpoint pen springs
- Fast-food wrappers
- Potato chip bag liners

You can also use aluminum from the following items:

- Margarine wrappers
- Ketchup and condiment packages
- Breath mint container labels
- Chewing gum wrappers
- Trading card packaging
- Coffee creamer container lids

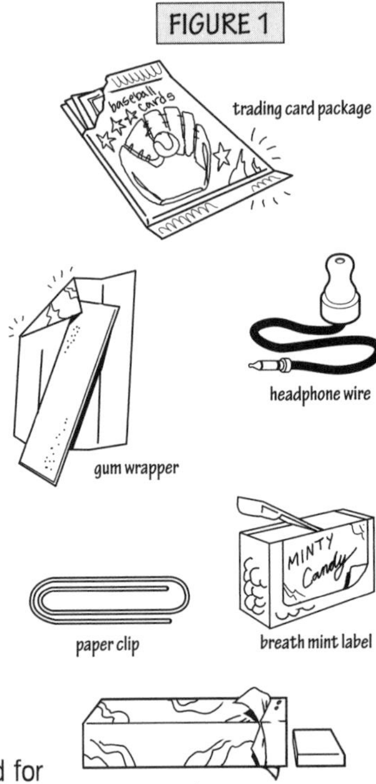

Note: The wire used from the sources above are only to be used for low-voltage, battery-powered projects.

More Power to You: Make Batteries from Everyday Things

No one can dispute the usefulness of electricity. But what do you do if you're in a remote area without AC power or batteries? Make sneaky batteries, of course!

In this project, you'll learn how to use fruits, vegetable juices, paper clips, and coins to generate electricity.

What's Needed

- Lemon or other fruit
- Nail
- Heavy copper wire
- Paper clip or twist-tie
- Water
- Salt
- Paper towel
- Pennies and nickels
- Plate

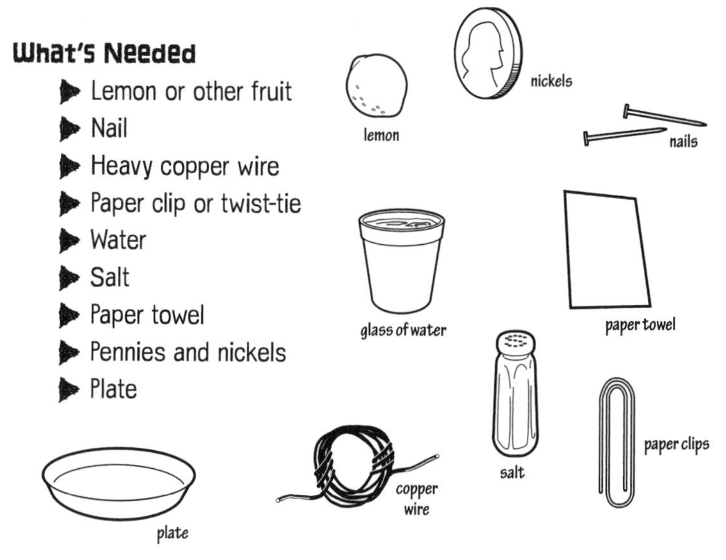

What to Do

The Fruit Battery

Insert a nail or paper clip into a lemon. Then stick a piece of heavy copper wire into the lemon. Make sure that the wire is close to, but does not touch, the nail (see **Figure 1**). The nail has become the battery's negative electrode and the copper wire is the positive electrode. The lemon juice, which is acidic, acts as the electrolyte. You can use other item pairs besides a paper clip and copper wire, as long as they are made of different metals.

The lemon battery will supply about one-fourth to one-third of a volt of electricity. To use a sneaky battery as the battery to power a small electrical device, like an LED light, you must connect a few of them in series, as shown in **Figure 2.** *Note:* If the LED does not light, reverse the connections on its leads.

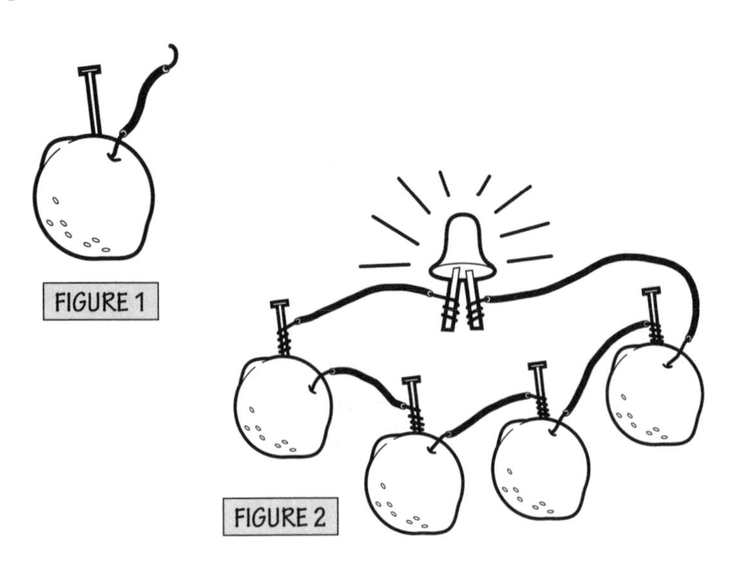

FIGURE 1

FIGURE 2

The Coin Battery

With the fruit battery, you stuck the metal into the fruit. You can also make a battery by placing a chemical solution between two coins.

Dissolve 2 tablespoons of salt in a glass of water. This is the electrolyte you will place between two dissimilar metal coins.

Now moisten a piece of paper towel or tissue in the salt water. Put a nickel on a plate and put a small piece of the wet absorbent paper on the nickel. Then place a penny on top of the paper. Next place another moistened piece of paper towel and then another nickel on top of the penny and continue the series until you have a stack.

In order for the homemade battery to do useful work, you must make a series of stacked coins and paper.

Be sure the paper separators do not touch one another.

The more pairs of coins you add, the higher the voltage output will be. One coin pair should produce about one-third of a volt. With six pairs stacked up, you should be able to power a small flashlight bulb, LED, or other device when the regular batteries have failed. See **Figure 3**. Power will last up to two hours.

Once you know how to make sneaky batteries, you'll never again be totally out of power sources.

FIGURE 3

Make Invisible Ink

If anything is a prime example of a Sneaky Use project, it's using everyday things to make invisible ink. (Sneaky fact: Casinos now use cards marked with symbols that are only visible when viewed with a special lens.) You can use a large variety of liquids to write secret messages. In fact, some prisoners of war used their own saliva and sweat to make invisible ink.

What's Needed

- Milk or lemon juice or equal parts baking soda and water
- Small bowl
- Cotton swab or toothpick
- Paper

What to Do

Use a cotton swab or toothpick to write a message on white paper, using the milk or lemon juice or baking soda solution as invisible ink. The writing will disappear when the "ink" dries.

To view the message, hold the paper up to a heat source, such as a lightbulb. The baking soda will cause the writing in the paper to turn brown. Lemon or lime juice contain carbon and, when heated, darken to make the message visible.

You can also reveal the message by painting over the baking soda solution on the paper with purple grape juice. The message will be bluish in color.

Sneaky Invisible Ink II

Here's another sneaky method to write and view invisible messages that stay invisible (unless you know the trick), this time using common laundry detergent and water.

What's Needed

- Cotton swab or towel
- Small bowl
- Liquid detergent
- Water
- Black light
- Piece of white cardboard

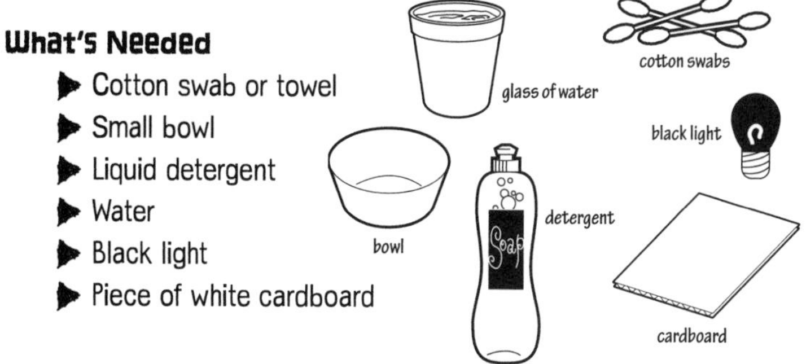

What to Do

In a small bowl, mix a teaspoon of liquid laundry detergent with one cup of water and use a small towel or cotton swab to write a message on a white piece of cardboard. See **Figure 1**. The message will not be visible at this point.

To view the secret message, darken the room and shine a black light—invisible ultraviolet light—on the board. The previously invisible message will become visible, as shown in **Figure 2**.

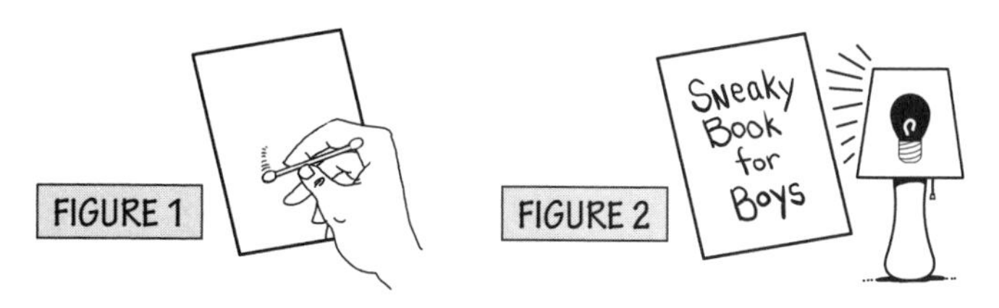

FIGURE 1 FIGURE 2

Part III

Sneaky Resourcefulness

You don't have to be MacGyver to adapt gadgets or survive in the wild. Anyone can learn how to be resourceful in minutes using nothing but a few items fate has put at your disposal. When you're in a bind, the best answer is frequently not the obvious one—it's the sneaky one.

Can you imagine being in an emergency without water, food, tools, or a way to signal for help? This section will illustrate how to trap small animals, lure fish, and construct a flotation device.

You will learn little-known methods to obtain water from the air, find your direction, craft a compass without a magnet, and more. Also included are sneaky projects that show how to modify toys and candy package accessories into useful alarms and security devices.

After learning the projects and techniques in this section, you will be ready to improvise simple security devices and safety gear in a pinch. You too can do more with less!

Emergency Signaling

Being stranded in a remote area can fill you with fear. What's especially frustrating is seeing a plane or vehicle and not being able to get their attention to be rescued.

This section will supply a few ways to signal for help. With a shiny object that reflects sunlight easily, you can signal to people and vehicles for assistance.

What's Needed

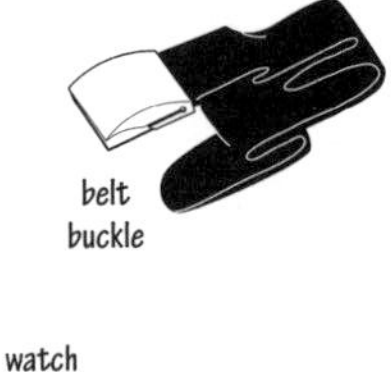

- Mirror, or belt buckle, metal pan or cup, or aluminum foil
- Reflective materials: canteen, watch, soda can, eyeglasses

What to Do

Using a mirror or other shiny object, point the light in one area and away in an SOS pattern (three short flashes, three long, and three short). Repeat this sequence of flashes as long as possible while sunlight is available until rescued (see **Figure 1**).

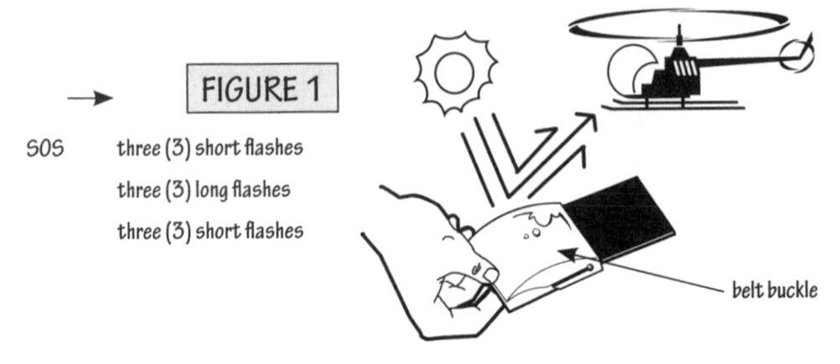

The *U.S. Army Survival Manual* recommends:

Do not flash a signal mirror rapidly, because a pilot may mistake the flashes for enemy.

Do not direct the beam in the aircraft's cockpit for more than a few seconds as it may blind the pilot.

Haze, ground fog, and mirages may make it hard for a pilot to spot signals from a flashing object. If possible, therefore, get to the highest point in your area when signaling. If you can't determine the aircraft's location, flash your signal in the direction of the aircraft noise.

At night you can use a flashlight or a strobe light to send an SOS to an aircraft.

Other Sneaky Signaling

When you're lost, use anything and everything as a marker to be seen by aircraft and search parties. Natural materials—snow, sand, rocks, vegetation—and clothing can be used as pointers to spell out distress signals. Follow the Ground-to-Air Emergency Code in laying out your markers.

symbol	message
I	Serious Injuries, Need Doctor
II	Need Medical Supplies
V	Require Assistance
F	Need Food and Water
LL	All Is Well

Y	Yes or Affirmative
N	No or Negative
X	Require Medical Assistance
–>	Proceeding in This Direction

body signal	**message**
Both arms raised with palms open	"I need help"
Lying on the ground with arms above head	"Urgent medical assistance needed"
Squatting with both arms pointing outward	"Land here"
One arm raised with palm open	"I do not need help"

See **Figure 2** for illustrations of these signals.

To show that your signal has been received and understood, an aircraft pilot will rock the aircraft from side to side (in daylight or moonlight) or will make green flashes with the plane's signal lamp (at night). If your signal is received but *not* understood, the aircraft will make a complete circle (in daylight or moonlight) or will make red flashes with its signal lamp (at night).

Sneaky Water-Gathering Techniques

In a survival situation, finding water is crucial; without it, you can only survive a few days. Drinking water from the ocean can be dangerous because of its 4-percent salt content. It takes about two quarts of body fluid to rid the body of one quart of seawater. Therefore, by drinking from the ocean, you deplete your body's water, which can lead to death.

Fresh drinking water can be gathered from a variety of sources. This project will show how to gather rainwater and dew from the air.

Collecting Dew

What's Needed

- Clean towel or cloth
- Cup, bowl, or other container

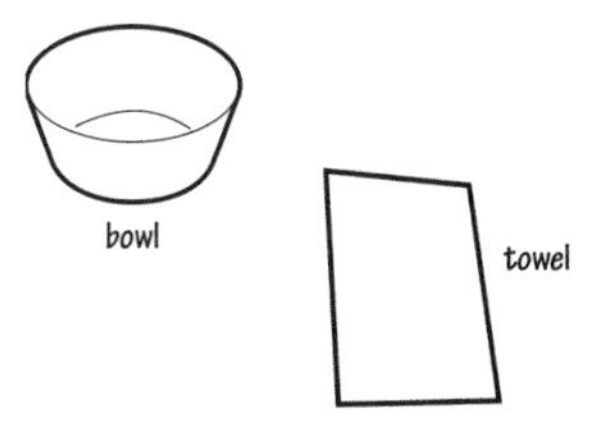

What to Do

In the early morning, dew forms on grass, plants, rocks, and other large surfaces near the ground because these items have cooled and water vapor condenses on their surface. The dew can be easily gathered by laying a clean towel on the dew-covered area, dampening it, and wringing the towel out over a bowl. See **Figure 1**.

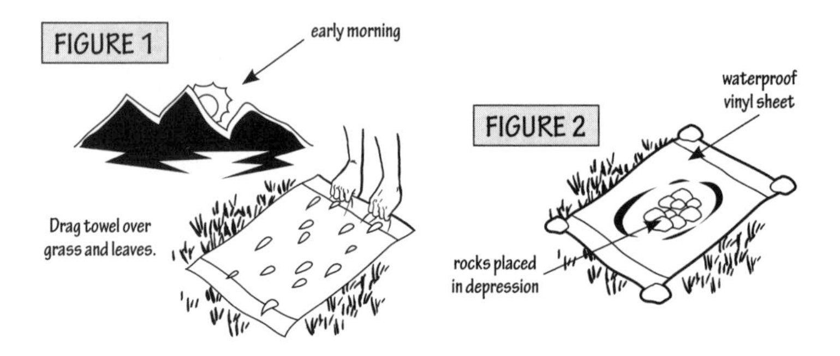

Gathering Rainwater

Rainwater, when available, is the preferred choice for drinking because it does not require boiling or purification. It can easily be collected by setting out items that you may already have.

What's Needed

- Cups, bowls, or other leakproof containers
- Plastic or vinyl material or a nylon jacket

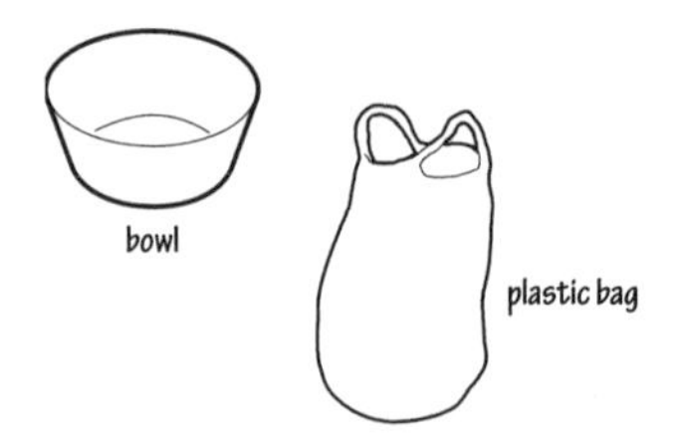

What to Do

Place all available cups and containers where they can fill with rainwater. If necessary, use waterproof material—plastic, vinyl, or a waterproof article of clothing—as a substitute container, as shown in **Figure 2.** Or make a container from a large leaf or from coated paper, as shown in the bonus application Make a Sneaky Cup in the next setting.

Get Drinking Water from Plants

Water is all around us in the air. The trick in obtaining it is to make it condense on the surface of an object and then collect it in a container.

Evaporator Still

An evaporator still can be made with a clear or translucent plastic bag and a large plant. It works by allowing the sun to shine through the bag and heat the plant, causing it to give off water vapor through its leaves. The water vapor condenses on the inside surface of the bag and drips down to the bottom. It can then be used for drinking water.

What's Needed

- Large plastic bag, preferably clear

What to Do

Gather green leaves or grass and place them in a plastic bag in a recessed area of the ground, as shown in **Figure 1**. Select an area where there will be plenty of sunlight. Or choose a plant or leafy tree branch, brush off any excess particles, wrap it in a plastic bag, and secure its opening with string, wire, or a tight knot; see **Figure 2**.

As the bag heats up, water from the leaves will evaporate and then condense in the bag as droplets that can be consumed later.

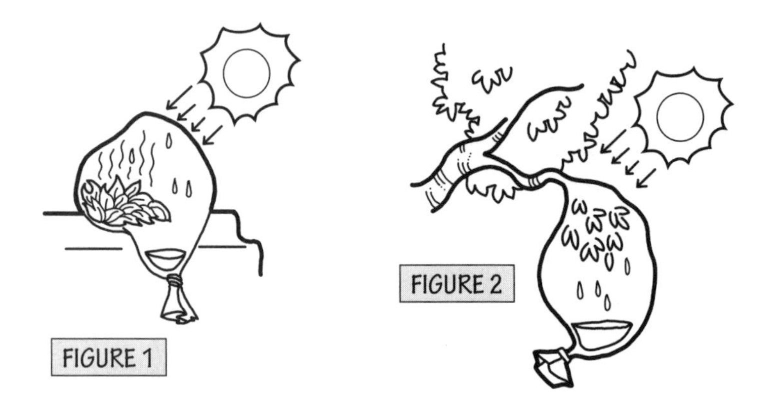

FIGURE 1

FIGURE 2

Overground Solar Still

You can survive up to a month without food but only a few days without water. In the wilderness, there's always a concern about obtaining fresh drinking water. If you are near vegetation and have a large plastic bag available, you can quickly construct a solar still to acquire water.

A solar still uses heat to draw moisture from air, ground, or plants. It then collects the moisture droplets and condenses them into a container for drinking. Solar stills are easy to make, but the amount of water they produce will vary depending on their size, the amount of sunlight, and the terrain.

What's Needed

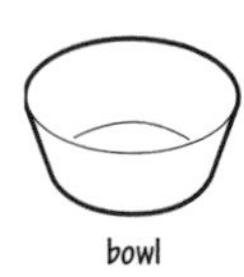

- Plastic bag, preferably clear, or plastic or vinyl material
- Cup or bowl or watertight container
- Rocks
- Stick
- Digging utensil

What to Do

The evaporator still proves that water from the air and from plant material can be trapped inside a plastic bag. With an overground solar still, you must dig out an oval or triangular trench and then another around it in an oval shape, as seen in **Figure 1**. Create the trench on an incline so that water will flow toward the end of the oval section.

First, place a tall stick in the center of the still and set the plant materials inside the center trench; see **Figure 2**. Next, cover both the stick and both trenches with the plastic bag and hold it in place with rocks. Last, ensure that the bag end is closed and secure. Water from the plants will heat up in the sun, evaporate, condense on the inner surface of the plastic bag, and run down the sides and into the closed end of the bag in the outer oval trench, where it can be poured into a container later; see **Figure 3**.

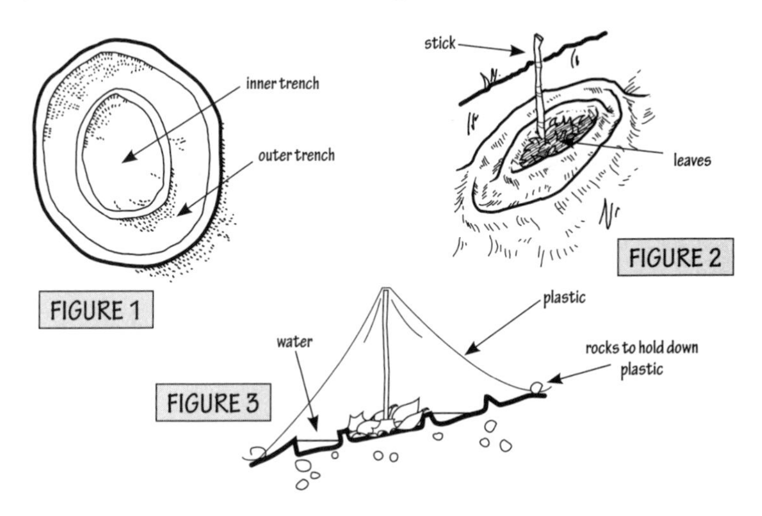

FIGURE 1

FIGURE 2

FIGURE 3

H2Origami—Make a Sneaky Cup

Gathering or extracting condensed water from the air will be futile without a bowl or cup. You can make a sneaky cup from paper—preferably coated paper from a magazine—or from a large leaf from a tree.

What to Do

The following illustrations show a piece of paper with two sides. Each step is shown in its corresponding illustration figure number.

1. Start with a square piece of paper or large leaf.
2. Fold corner B diagonally on top of corner C.
3. Fold corner A, now point A, down as shown.
4. Fold corner D, now point D, to the opposite edge, to the place where the first crease hits the edge of the paper.

5. Fold the paper on these two creases.
6. Fold the front (top) flap, corner B, down to cover all the layers. Fold the other flap, corner C, backward (there is only one layer to cover in the back).

7. Open the cup by pulling the front and the back apart.

The sneaky cup can be placed underneath a dripping condensation gathering area to save fresh water.

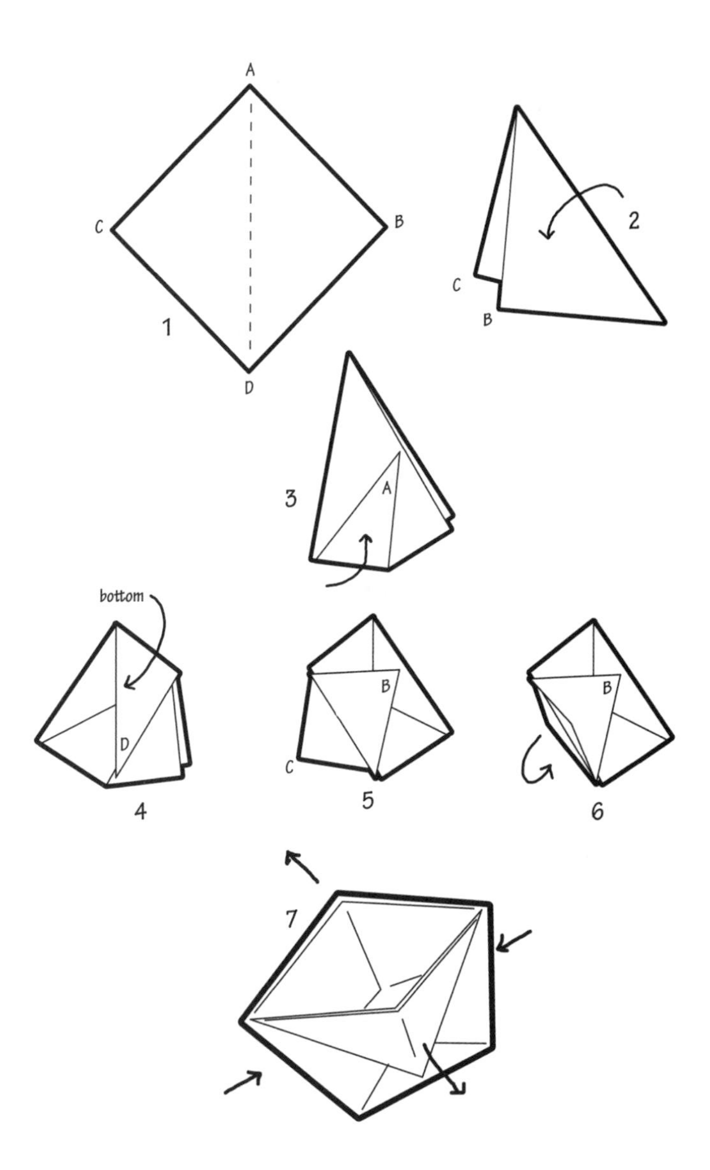
A
C
B
1
D
2
C
B
3
A
bottom
D
4
B
C
5
B
6
7

Sneaky Animal and Fish Traps

If you're ever trapped in the wilderness, capturing small animals for food can prove to be energy-wasting and very difficult. Using items you may already have in your pocket, along with rope, twigs, string, twine, and wire, you'll be able to set small sneaky traps to catch emergency food.

By setting several traps, you'll have the ability to catch much more potential food than by chasing animals with a weapon (which you probably don't have anyway!).

What's Needed

- Box
- Sticks
- Wire
- Strong threads from clothing or string
- Belt
- Vines
- Tree branches
- Small bits of food or worms
- Bottle
- Small rocks
- Large rock

What to Do

Box Trap

First, locate a cardboard box or use wood branches tied together with vines or strings to make a sneaky box shape. The box must have a door that can be propped open with a small branch in order to close behind the animal.

Next, position some small food or worms or a shiny item to lure the animal into the box opening. Set the stick so that it is positioned to keep the box door open gently, not rigidly. This allows the animal to bump into it and inadvertently close the door, thereby trapping the animal inside as shown in **Figure 1**.

If the box has an open top or no door flap, place the branch so that it props the box up off the ground. Set the bait near the base of the branch so when the curious animal moves about, it will bump into the stick and cause the box to drop on him. See **Figure 2**.

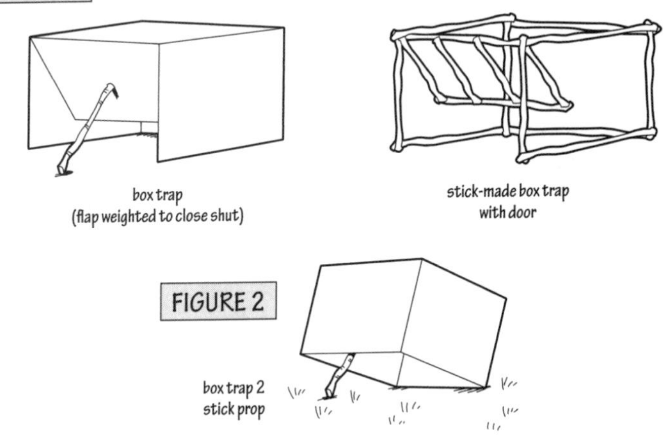

Snare Trap

First, locate items like long vines, thread from clothing, string, twine, or wire to make into a sneaky rope. Braid the material for strength and tie knots in both ends.

Next, tie one end of the rope around the base of a tree and secure it with a knot. Make a loop with the other end; this will become the snare. See **Figure 3.**

Hide the loop with grass and twigs placed on top of it. Place small bits of food or worms to attract an animal into the snare's loop. When a small animal runs into the loop, it will attempt to escape and pull it tighter, thus becoming trapped more tightly.

Rock Trap

If no box is available, use a large rock to trap a small animal. As for the box trap, place a branch so it props up the rock. Position the stick very delicately, because it should fall at the lightest touch. Place bait material next to the base of the branch so that the animal will knock over the stick while investigating and become trapped by the weight of the rock, as shown in **Figure 4.**

Sneaky Fishing

Experienced outdoorsmen know the value of being able to lure and trap fish. In an emergency survival situation, when you're near water and cannot get back to your campsite, you can use everyday items to make sneaky fishing rigs. A sneaky fishing rig consists of a line, a hook, and a lure.

The line extends the hook and lure deep into the water. The lure attracts the fish, and, with a little luck, the hook ensnares it, allowing you to pull it in for a meal. If the hook and lure stay too close to the water's surface, tie a small rock to the line so it will sink deeper into the water.

What's Needed

- Foil or other shiny object
- Bait (worms, insects, food bits, or feathers)
- Wire
- Strong threads from clothes or string
- Dental floss
- Paper clip
- Vine
- Plastic bottle
- Tree branches

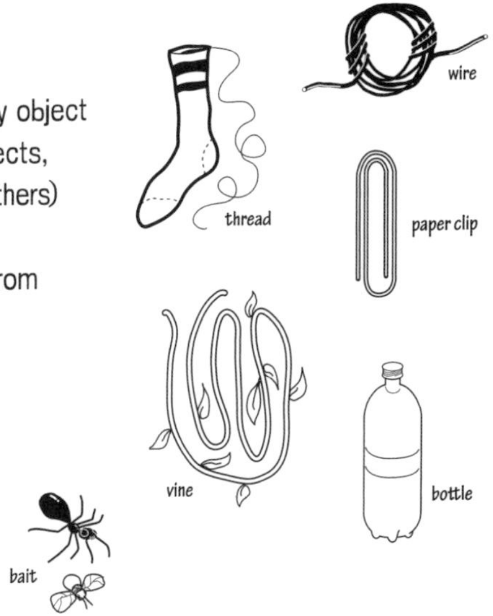

What to Do

Fishing Rigs

Lines. Fashion your fishing line from items you have on or around you, such as dental floss, clothing thread, or wire. Nature provides tree vines that, when flattened with a rock and braided together, can make a good makeshift fishing line.

Hooks. You can make sneaky fishhooks using pins, needles, wire, small nails—even a thorny branch. Paper clips, a straightened key ring, stiff wire, shells, and bones can also be used.

Lures. Just because a hook is dangling from a line, you can't depend on a fish to investigate it, so some sort of bait is required. Use whatever you can find in your area that can be stuck on the end of the hook. Insects, worms, and small bits of food will do the trick. If you're out of real food, objects like a button, a shiny chain or foil wrapper, a feather, a small key, or even a fish-shaped leaf will increase the odds that you'll lure a fish to swallow the hook.

Simply wrap the line around the hook, place the bait item on the hook, and extend the line as far out and into the water as you can as shown in **Figure 1**. If possible, prop up the line with a forked branch to allow it to extend farther into the water.

FIGURE 1

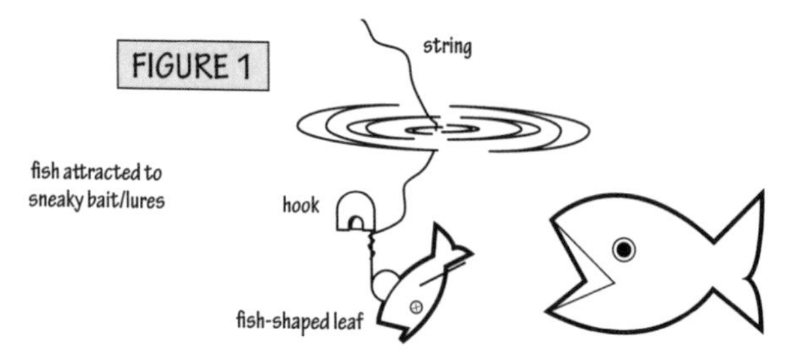

Fish Nets

If you're near a stream, you may be able to catch fish with a net, which is easy enough to make.

First, find a long tree branch that splits off into a fork and remove all the leaves on the branch. Take a spare shirt and tie the sleeves and collar area into a knot.

Then place the shirt upside down into the forked branch so it produces a makeshift net. Secure the shirt to the branch using whatever you have—paper clips, a key ring, wire, or smaller forked branches—to keep the shirt tight on the branch ends. See **Figure 2**.

Last, lay the branch and shirt into the water, and you can ensnare aquatic creatures in your Sneaky Net.

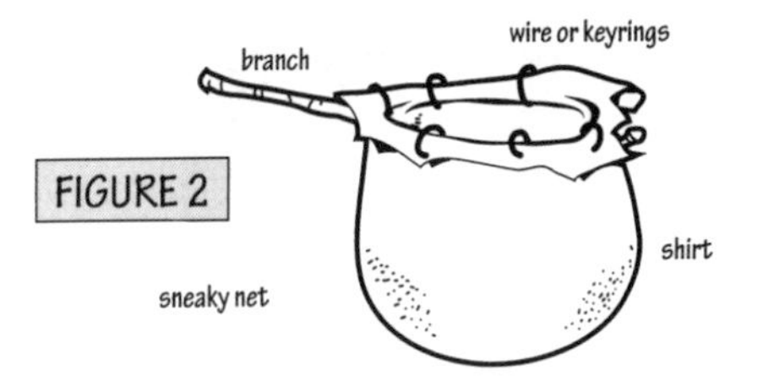

Aqua Traps

A spare plastic bottle can provide another sneaky method to trap small fish if you're near a stream or pond.

First, use a knife or sharp rock to cut off the top of the bottle. Then place some sort of bait material in the bottom of the bottle. You can use insects, worms, small food bits, a button, a shiny chain, a foil wrapper, or a feather.

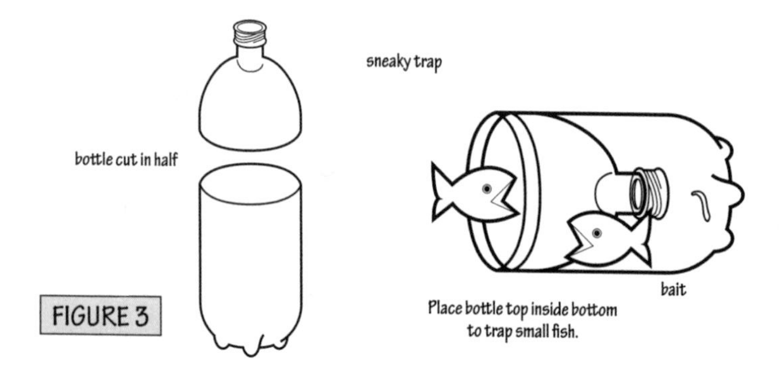

FIGURE 3

Next, turn the bottle top upside down and push it, mouth down, into the bottom section so it's wedged tight. See **Figure 3**. Now place the bottle near the edge of a stream or pond. Curious fish will swim through the mouth of the bottle looking for the bait but will not be able to get out. Leave as many of these aqua traps as you can in the water, and with a little luck you will find a fish dinner waiting for you on your return.

Emergency Flotation Devices

If you find yourself in a sink-or-swim scenario, what will you do if a flotation device isn't available? Make a sneaky one from everyday things.

When floating in water, the more you try to keep your head above the surface of the water, the more likely you are to sink. Just lie back and keep your mouth above water.

When you attempt to raise parts of your body above the surface, you lose buoyancy. Luckily, however, you can add to your buoyancy with virtually any empty container that holds air. In some instances two or more may need to be secured together.

What's Needed

- Plastic bags
- Gas cans
- Large soda bottles
- Other items that will hold air
- String, wire, a belt, or cloth

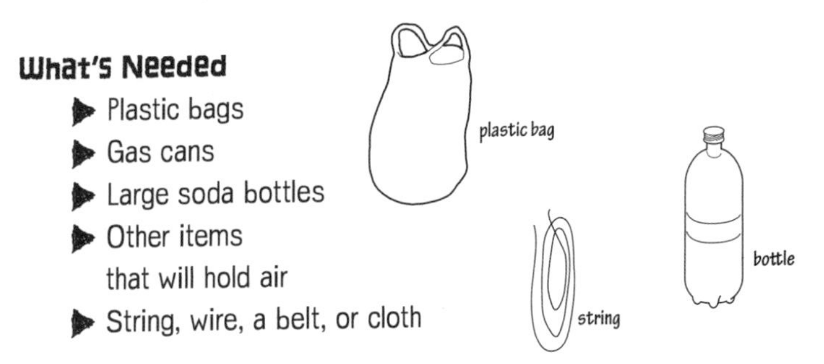

What to Do

To make a flotation device of plastic bags, blow the smallest one up, tie a tight knot, and place it in a larger bag (or bags, if available) as shown in **Figure 1** to compensate for small holes. Use these inflated bags as water wings to help stay afloat. Rest on your back with your head up (**Figures 2** and **3**).

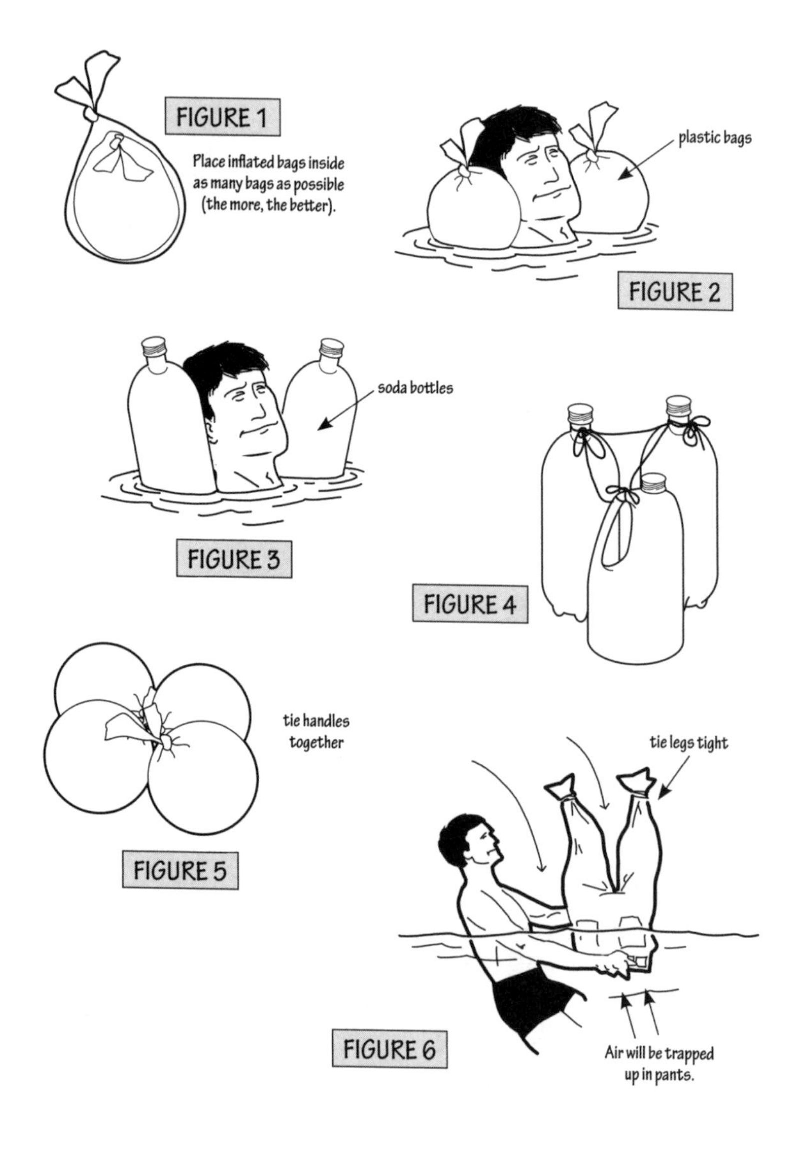
FIGURE 1
Place inflated bags inside as many bags as possible (the more, the better).
plastic bags
FIGURE 2
soda bottles
FIGURE 3
FIGURE 4
tie handles together
FIGURE 5
tie legs tight
FIGURE 6
Air will be trapped up in pants.

Figure 4 illustrates how to connect two water-holding bottles or jars together. You can also use a log if one is available. Be sure the log will float before laying your body on it (not all wood will float).

There are many other flotation devices that you can devise by using some imagination. Just make sure to test their flotation capabilities before trying to use them. **Figure 5** shows how to tie bags together.

When no other items are available, your clothing can hold pockets of air to increase your buoyancy. If necessary, remove your pants or shirt, tie the ends of the pant legs or sleeves in knots and scoop air into them. Hold the other end together firmly with your hands and you should be able to ride above the water with little effort (see **Figure 6**).

Ink or Swim

Who among us has not feared for our small companions, children and pets alike, when they're near a body of water?

With a few everyday items, you can prevent a possible drowning. Just bring this easy-to-make lifesaving gadget with you when traveling over or near water with your little ones.

What's Needed

- Large wide-barreled pen
- Four large balloons (two for each arm)
- Four large rubber bands

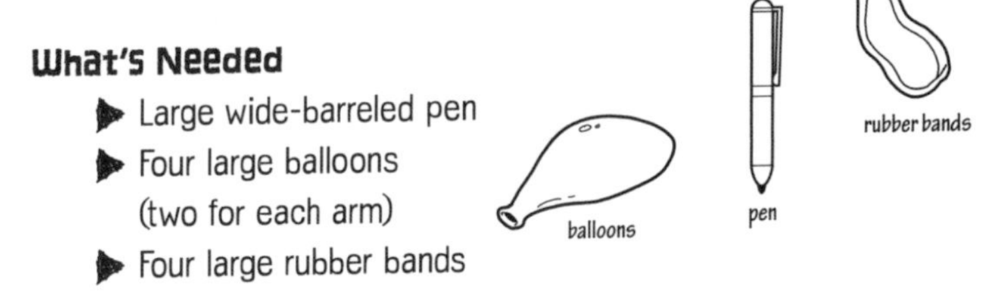

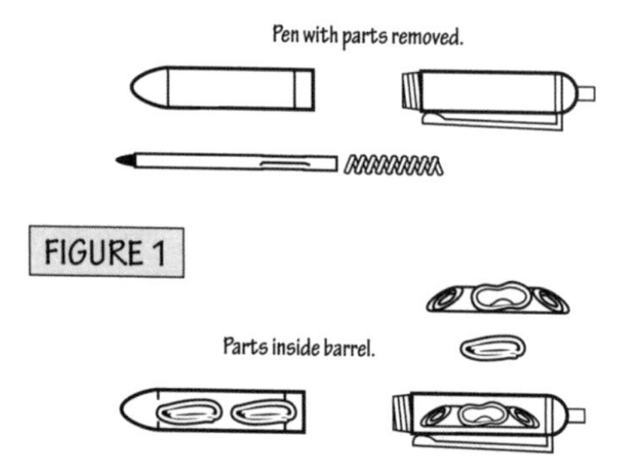

What to Do

You can store emergency sneaky water wings available for nonswimmers inside a large hollow pen barrel. When tightly wrapped, the balloons and rubber bands will easily fit inside a pen, as shown in **Figure 1**. Just slip the pen in a pocket or purse before a trip for added peace of mind.

Figure 2 shows how to affix the sneaky water wings to a child's arms in case of an emergency swim. If the rubber bands are too large for the arms or legs of the child (or pet), just wrap them around multiple times for a snug fit.

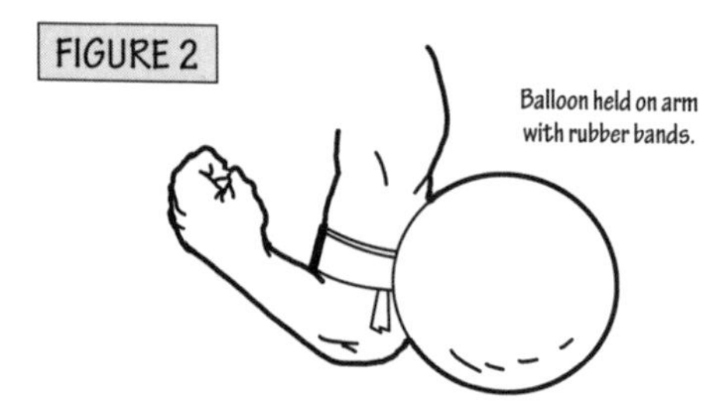

Craft a Compass

If you're ever lost, you'll find a compass is a crucial tool. When markers or trails are nonexistent, a compass can keep you pointed in the right direction to get you back to a line of reference.

A compass indicates Earth's magnetic north and south poles. For a situation where you are stranded without a compass, this project describes three ways of making one with the things around you. For each method, you will need a needle (or twist-tie, staple, steel baling wire, or paper clip); a small bowl, cup, or other nonmagnetic container; water; and a leaf or blade of grass. How simple is that?

Method 1

What's Needed

- Magnet—from a radio or car stereo speaker

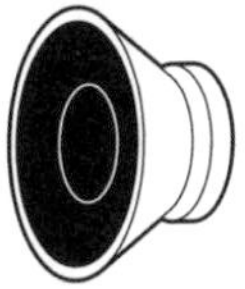

stereo speaker

What to Do

Take a small straight piece of metal (but do not use aluminum or yellow metals), such as a needle, twist-tie, staple, or paper clip, and stroke it in one direction with a small magnet. Stroke it at least fifty times, as shown in **Figure 1**. This will magnetize the needle so it will be attracted to Earth's north and south magnetic poles.

Fill a bowl or cup with water and place a small blade of grass or any small article that floats on the surface of the water. Place the needle on the blade of grass (see **Figure 2**) and watch it eventually turn in one direction. Mark one end of the needle so that magnetic north is determined.

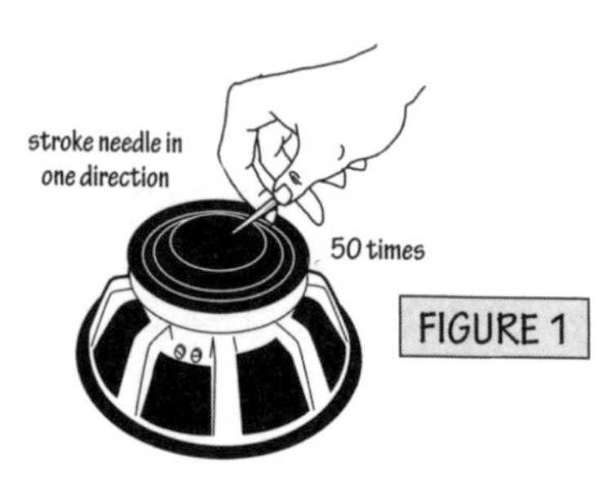

FIGURE 1

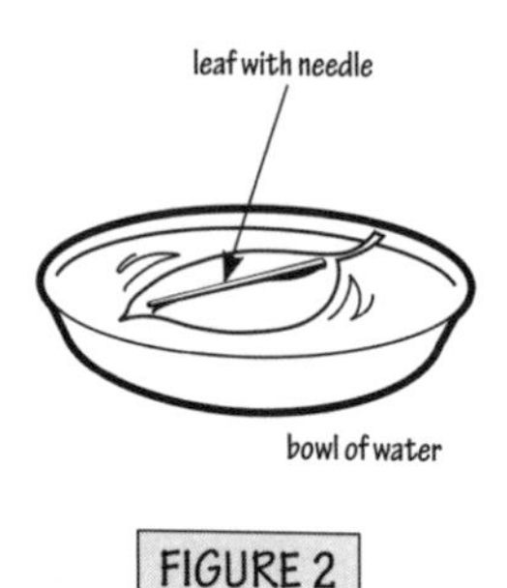

FIGURE 2

Method 2

What's Needed

- Silk or synthetic fabric—
  from a tie, scarf, or other garment

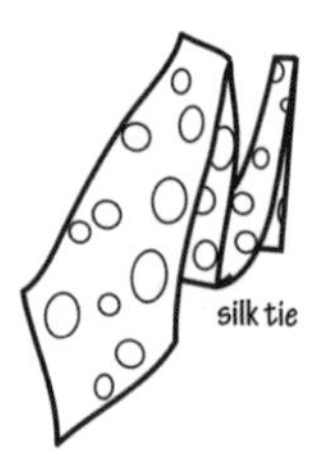

What to Do

As in the first method, stroke a needle or paper clip in one direction with the silk material. This will create a static charge in the metal, but it will take many more strokes to magnetize it. Stroke at least 300 times, as shown in **Figure 3**. Once floated on a leaf in the bowl, the needle should be magnetized enough to be attracted to Earth's north and south magnetic poles. You may have to remagnetize the sneaky compass needle occasionally.

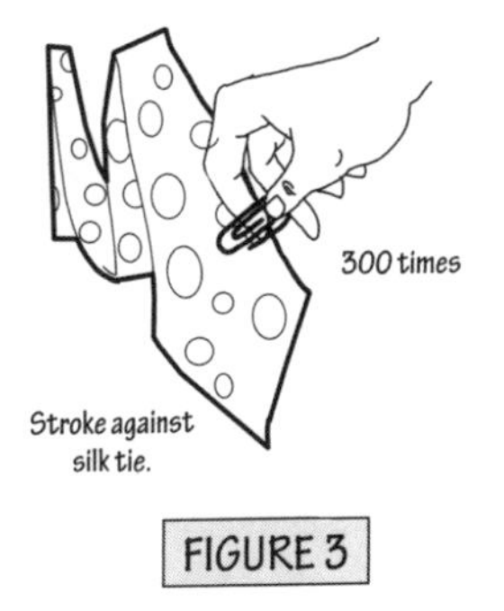

FIGURE 3

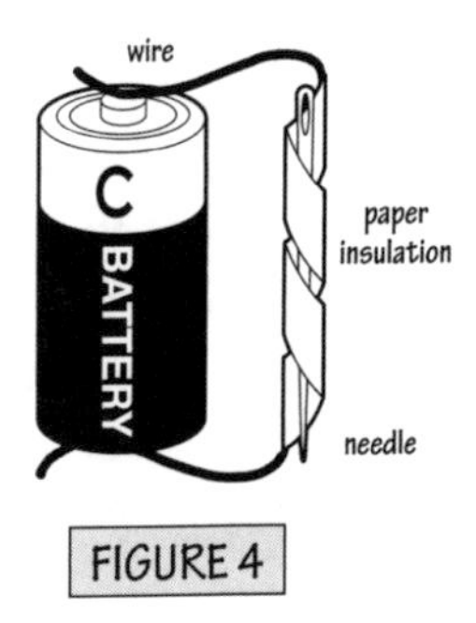

FIGURE 4

Method 3

What's Needed

- Battery

What to Do

When electricity flows through a wire, it creates a magnetic field. If a small piece of metal, like a staple, is placed in a coil of wire, it will become magnetized.

Wrap a small length of wire around a staple or paper clip and connect its ends to a battery, as shown in **Figure 4**. (For sneaky battery and wire sources ideas, see "Sneaky Science Projects" in Part II.) If the wire is not insulated, wrap the staple with paper or a leaf and then wrap the wire around it.

When you connect the wire to the battery in this manner, you are creating a short circuit—an electrical circuit with no current-draining load on it. This will cause the wire to heat quickly so only connect the wire ends to the battery for short four-second intervals. Perform this procedure fifteen times.

Place the staple on a floating item in a bowl of water, and it will eventually turn in one direction. Mark one end of the staple so that magnetic north is determined.

Direction Finding Methods

If you're stranded without a magnetic compass, all is not lost. Even without a compass, there are numerous ways to find directions in desolate areas. Three methods are covered here.

Method 1: Use a Watch

What's Needed

- Standard analog watch
- Clear day where you can see the sun

What to Do

The sun always rises in the east and sets in the west. You can use this fact to find north and south with a standard nondigital watch.
If you are in the Northern Hemisphere (north of the equator), point the hour hand of the watch in the direction of the sun. Midway between the hour hand and 12 o'clock will be south. See **Figure 1**.

FIGURE 1

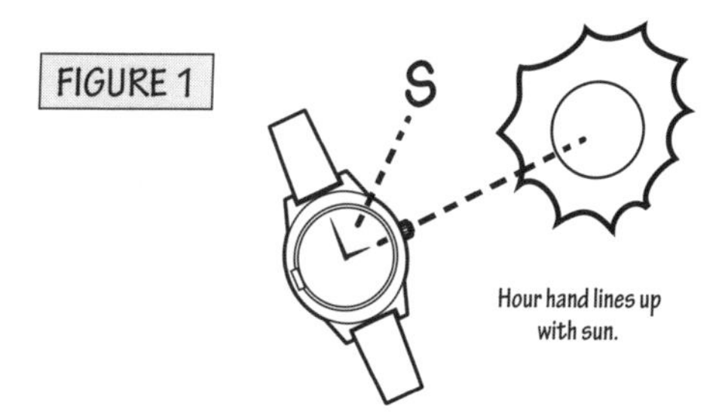

Method 2: Use The Stars

What's Needed

- A clear evening when stars can be viewed

What to Do

In the Northern Hemisphere, locate the Big Dipper constellation in the sky; see **Figure 2**. Follow the direction of the two stars that make up the front of the dipper to the North Star. (It is about four times the distance between the two stars that make up the front of the dipper.) Then follow the path of the North Star down to the ground. This direction is north.

In the Southern Hemisphere, locate the Southern Cross constellation in the sky; see **Figure 3**. Also notice the two stars below the Cross. Imagine two lines extending at right angles, one from a point midway between the two stars and the other from the Cross, to see where they intersect. Follow this path down to the ground. This direction is due south.

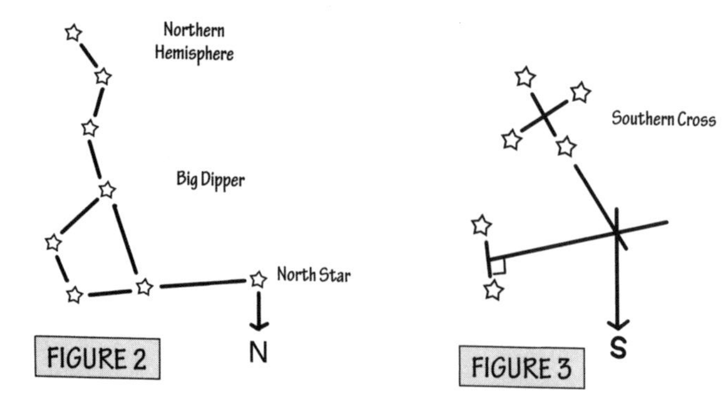

FIGURE 2

FIGURE 3

Method 3: Use a Stick

What's Needed

- Stick or branch about 3 feet long
- Rock or leaf

What to Do

On a sunny day, you can find out which direction is north, south, east, or west by using shadows. Stand a stick upright in the ground, as shown in **Figure 4**. Notice the shadow it casts and, using a rock or leaf, mark the shadow's edge.

Wait about fifteen minutes and notice the new shadow that appears. Mark its tip, too. See **Figure 5**. Draw an imaginary line between the two marks. This is the east–west line (west is the first tip, and the second marker represents east). You can draw an imaginary or real line across the east–west line to determine the north and south directions. See **Figure 6**.

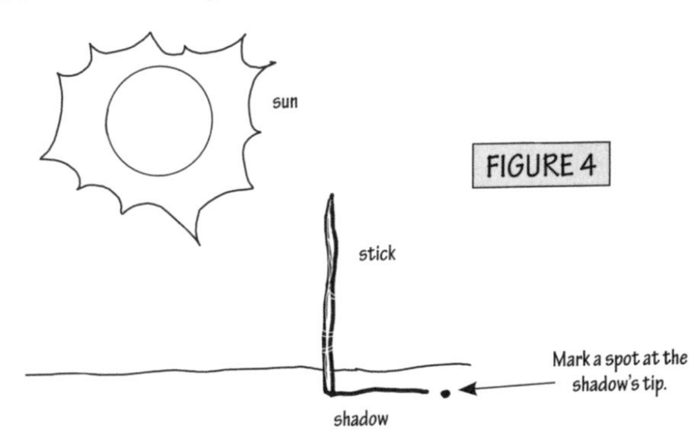

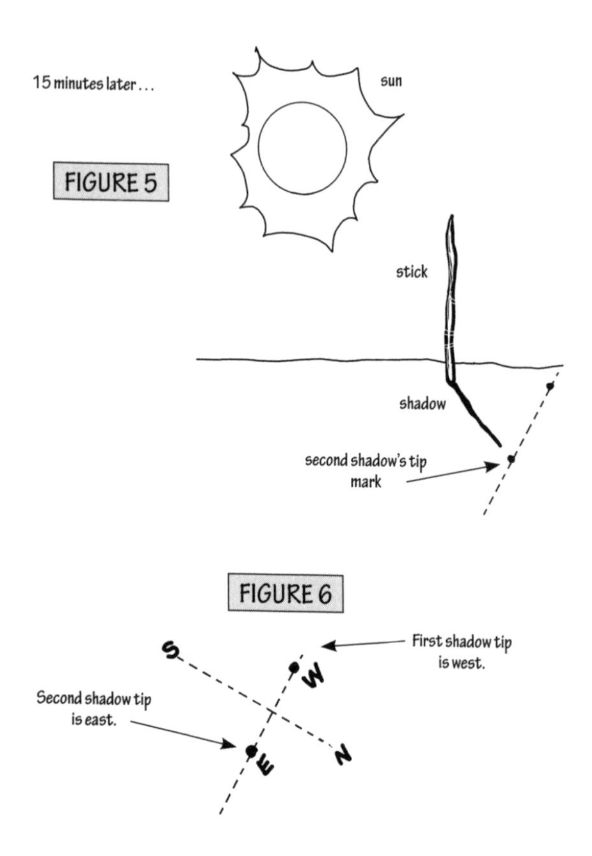
15 minutes later . . .
sun
FIGURE 5
stick
shadow
second shadow's tip mark
FIGURE 6
S
W
First shadow tip is west.
Second shadow tip is east.
E
N

Latitude and Longitude Info

Latitude

Latitude represents how far north or south you are, relative to the equator. When you're at the equator, your latitude is zero. The North Pole's latitude is 90 degrees north. Conversely, the South Pole is 90 degrees south.

Note: South latitude figures, which are south of the equator, are represented as negative numbers.

Longitude

Longitude represents how far east or west you are, relative to the Greenwich meridian. Places to the west of Greenwich have longitude angles up to 180 degrees west. Positions east of Greenwich have longitude angles up to 180 degrees east. Longitude west figures are input as negative numbers.

Decimal and Degrees/Minutes/Seconds Notation

Maps and GPS receivers show latitude and longitude angles. Maps usually show bold lines marked in degrees (whole numbers) plus possibly intermediate lines marked 15, 30, 45 minutes or 10, 20, 30,

40, 50 minutes. GPS receivers typically show degrees plus minutes and decimal fractions of a minute (e.g., 45 : 23.1234).

Each degree can be subdivided into 60 minutes (and each minute into 60 seconds for very high precision).

In cases where the map (or GPS readout) is in degrees and minutes, convert the minutes to decimals of a degree by dividing the number of minutes by 60. **For example:**

50 deg 30 minutes north = 50.5 degrees

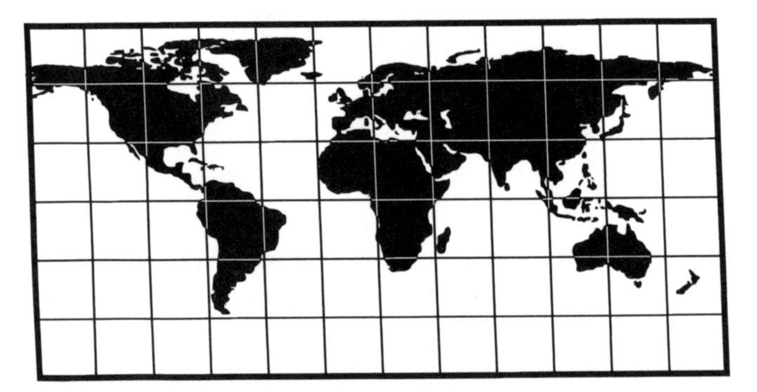

For example: Anchorage, Alaska, 61 13 149 54 = 61 degrees, 13 minutes north and 149 degrees, 54 minutes west. Degrees are sometimes represented with a ° symbol and minutes with a ' symbol. A typical notation looks like this: Anchorage, Alaska 61° 13' N 149° 54' W.

You can design and play a city location quiz by using a world map and listing only the coordinates from the list below—be sure to leave out the city names. Place the proper city names in an answer key to make a sneaky quiz.

For example, here is a sample quiz you can try:

 Which city is 41° 50' N 87° 37' W at 11:00 A.M.?

2. Which city is 42° 21' N 71° 5' W at 12:00 noon?

3. Which city is 40° 47' N 73° 58' W at 12:00 noon?

4. Which city is 33° 45' N 84° 23' W at 12:00 noon?

Answers: 1. Chicago, Ill., 2. Boston, Mass., 3. New York, NY., 4. Atlanta, Ga.

(**Example:** Chicago, Ill. 41° 50' N 87° 37' W at 11:00 A.M.
Notice that all of the cities below are degrees north and degrees west because they are all in North America.)

Cities	**Latitude**	**Longitude** (decimal and degrees/minutes)	**Time**
Anchorage, Alaska	61° 13' N	149° 54' W	8:00 A.M.
Atlanta, Ga.	33° 45' N	84° 23' W	12:00 noon
Baltimore, Md.	39° 18' N	76° 38' W	12:00 noon
Boise, Idaho	43° 36' N	116° 13' W	10:00 A.M.
Boston, Mass.	42° 21' N	71° 5' W	12:00 noon
Buffalo, N.Y.	42° 55' N	78° 50' W	12:00 noon
Chicago, Ill.	41° 50' N	87° 37' W	11:00 A.M.
Cincinnati, Ohio	39° 8' N	84° 30' W	12:00 noon

Dallas, Tex.	32° 46' N	96° 46' W	11:00 A.M.
Denver, Colo.	39° 45' N	105° 0' W	10:00 A.M.
Fargo, N.Dak.	46° 52' N	96° 48' W	11:00 A.M.
Honolulu, Hawaii	21° 18' N	157° 50' W	7:00 A.M.
Kansas City, Mo.	39° 6' N	94° 35' W	11:00 A.M.
Las Vegas, Nev.	36° 10' N	115° 12' W	9:00 A.M.
Los Angeles, Calif.	34° 3' N	118° 15' W	9:00 A.M.
Memphis, Tenn.	35° 9' N	90° 3' W	11:00 A.M.
Miami, Fla.	25° 46' N	80° 12' W	12:00 noon
Milwaukee, Wis.	43° 2' N	87° 55' W	11:00 A.M.
Minneapolis, Minn.	44° 59' N	93° 14' W	11:00 A.M.
New Orleans, La.	29° 57' N	90° 4' W	11:00 A.M.
New York, N.Y.	40° 47' N	73° 58' W	12:00 noon
Philadelphia, Pa.	39° 57' N	75° 10' W	12:00 noon
Pittsburgh, Pa.	40° 27' N	79° 57' W	12:00 noon
St. Louis, Mo.	38° 35' N	90° 12' W	11:00 A.M.
Salt Lake City, Utah	40° 46' N	111° 54' W	10:00 A.M.
San Francisco, Calif.	37° 47' N	122° 26' W	9:00 A.M.
Washington, D.C.	38° 53' N	77° 02' W	12:00 noon
Winnipeg, MB, Canada	49° 54' N	97° 7' W	11:00 A.M.

Sneaky Toy Modifications

Who among us hasn't dreamed of having a power door opener as seen in sci-fi and spy movies? This project will show you how to use a small toy car to do the trick. A small wire-controlled car has enough power to push and pull a typical room door back and forth if you know the super-sneaky way to install it.

What's Needed

- Wire-controlled toy car
- Velcro tape, adhesive-backed
- Screwdriver
- Pliers

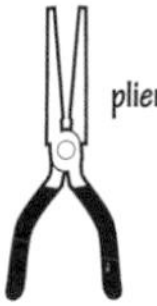

pliers

toy car

What to Do

This project requires a small wire-controlled toy car, not a radio-controlled version. This is to prevent the batteries from running down. (With a radio-controlled car, the remote control and the car's internal receiver have to be in the on mode, and this drains batteries.)

First, remove the body shell from the toy car with a screwdriver. Then remove the front wheel and axle, as shown in **Figure 1**. Now, using the Velcro tape, attach the car near the bottom end of the door (see **Figure 2**).

Using the remote control, see if it can push the door open or closed. If not, reposition the car for more traction. When the proper position is found, you will be able either to move the door with your hand or let the car do it.

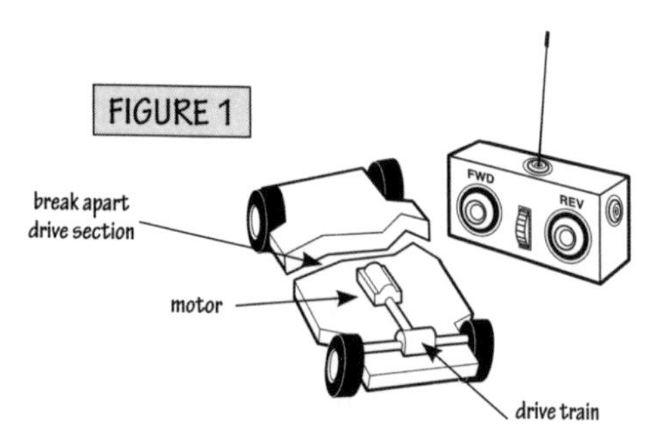

Optionally, you can break off the entire front part of the chassis so that it takes up less space and cover it with materials for a more appealing look. Mount the remote control outside the door as desired (see **Figure 3**).

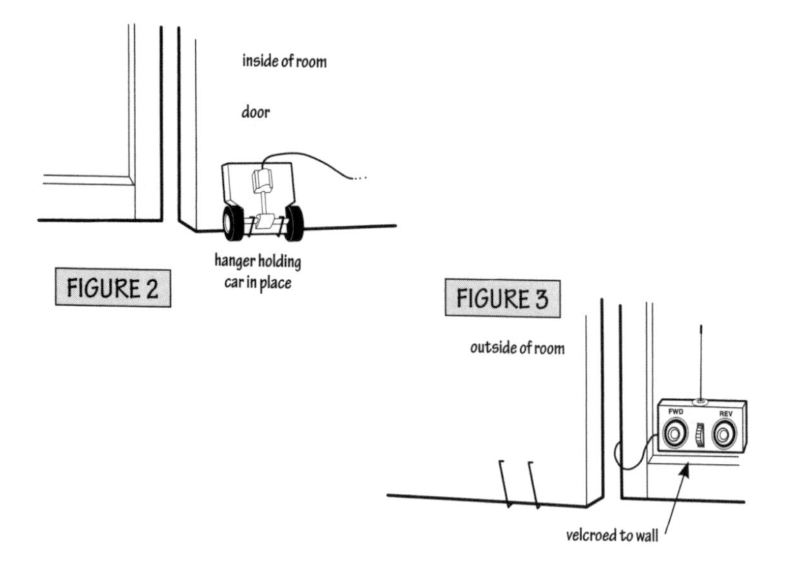

Sneaky Radio-Control Car Projects

Radio-controlled cars have many sneaky adaptation possibilities that can increase their usefulness. This project uses the inexpensive single-function type of radio-controlled toy car; this model will travel forward continuously, once its on/off switch is placed in the on position, until you actuate the remote control button, causing it to back up and turn. When you release the control, the vehicle goes forward in a straight line again.

The instructions and illustrations that follow will show you how to modify the transmitter to a more compact size, to use it as an alarm trigger. You'll also see how to modify the receiver to activate other devices, such as lights and buzzers.

What's Needed

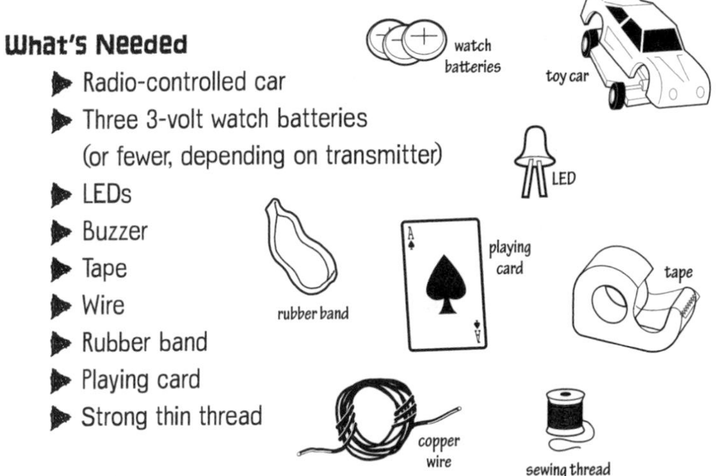

- Radio-controlled car
- Three 3-volt watch batteries (or fewer, depending on transmitter)
- LEDs
- Buzzer
- Tape
- Wire
- Rubber band
- Playing card
- Strong thin thread

What to Do

How a radio-controlled car works. Pressing the transmitter button closes an electrical switch, which turns on the transmitter. This sends electro-

FIGURE 1

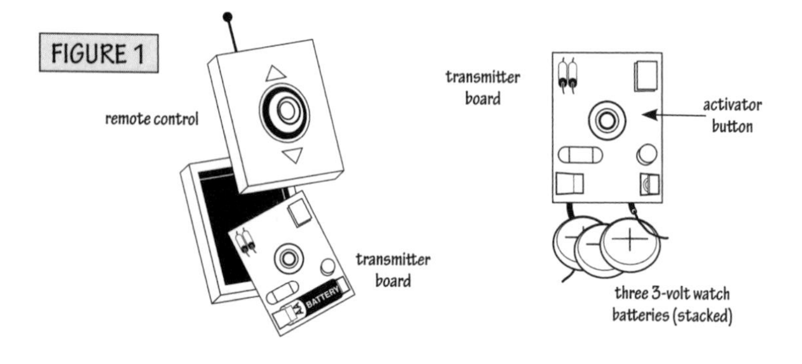

magnetic waves through the air that are detected by the radio receiver in the vehicle. The receiver detects the radio signal from the transmitter and reverses the electrical polarity (direction of current flow) of the power applied to the motor. This causes it to run in the reverse direction.

Adapting the car's transmitter. The first sneaky adaptation to the transmitter is to make it as small as possible, for concealment inside other objects or clothing.

Since the transmitter is always in the off mode until its activator button is pressed, it can operate using tiny long-life watch batteries.

If the transmitter uses one AA or AAA battery, it can be replaced by one small watch battery with the same voltage output. *Note:* Each AA or AAA battery supplies 1½ volts of power.

If the transmitter operates on two AA or AAA batteries, you can substitute either two 1½-volt watch batteries or a single 3-volt watch battery. If a 9-volt battery was in use, you will need to use three 3-volt watch batteries. When stacking batteries, place the positive side of one battery against the negative side of the other.

Figure 1 shows how to replace regular AA or 9-volt batteries in the transmitter with 3-volt watch batteries.

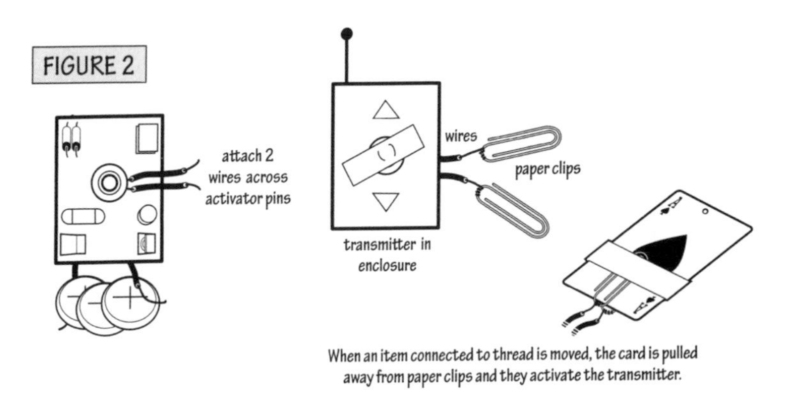

When an item connected to thread is moved, the card is pulled away from paper clips and they activate the transmitter.

If you connect two wires across the transmitter's activator button, you can have another sensor or switch activate the transmitter to alert you of an entry breech or that your valuables are being removed. Place a piece of tape over the transmitter button so that when the device is activated it will be on. See **Figure 2.**

Adaptating the car's receiver. You can also modify the car's radio receiver, which is on a circuit board in the car's body, for use as an alarm trigger. See **Figure 3.**

Unlike the transmitter, the receiver must stay on to be able to operate, and this produces a small constant drain on the batteries. **Figure 4** illustrates how to modify the receiver for use with watch batteries using the same technique described for the radio transmitter.

If desired, the toy car motor can be used in an application of your own design. The car motor is attached to the receiver with two connecting wires. If you physically remove the motor from the car body (either by unclipping or unscrewing it), you can use the receiver for more project applications. It's easy to connect the receiver's motor wires to other devices to activate them remotely.

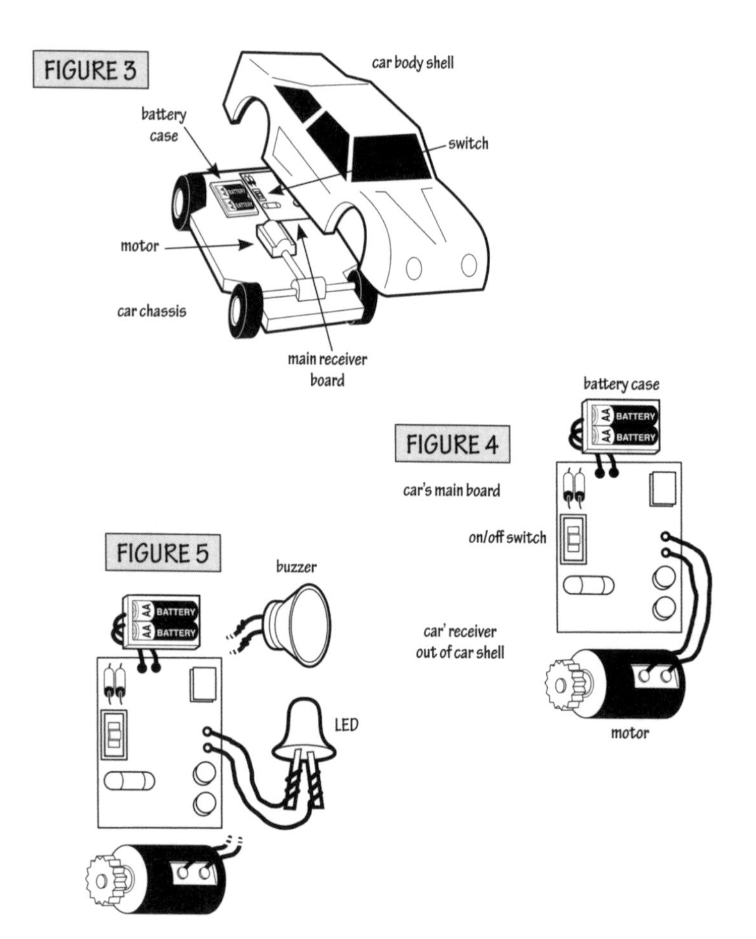

Figure 5 shows how the wires in the receiver that previously connected to the motor can be connected to other devices, like an LED or a buzzer for remote control.

Sneaky Walkie-Talkies

A pair of compact walkie-talkies lends itself to a variety of sneaky applications. Some models are so small, they are now mounted in toy wristwatches. This project illustrates methods to use walkie-talkies as an intercom, as an alarm sensor, and as a sneaky listening device.

What's Needed

- Pair of walkie-talkies
- Tape
- Nylon thread
- Coated business card or bookmark
- Nine-volt battery eliminator (optional)

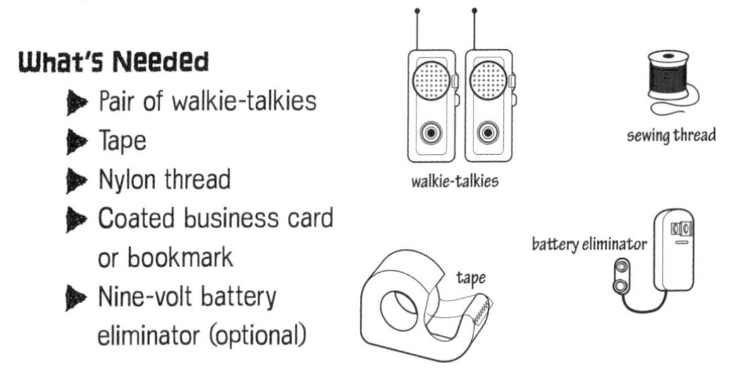

What to Do

First, test the walkie-talkies with fresh batteries and note their maximum reliable operating distance. Follow the directions in the categories below for your desired application.

Sneaky Intercom

This is the easiest project application to set up. It takes the place of an intercom system when wires cannot be placed between locations. For example, you can place one walkie-talkie in one room of the house and the other in the basement, the garage, a bedroom, or at the front door (mounted securely with screws or glue underneath a protective awning).

Simply mount the first walkie-talkie in a remote area (possibly under a cover to protect it from the elements) and listen with the second unit as shown in **Figure 1**.

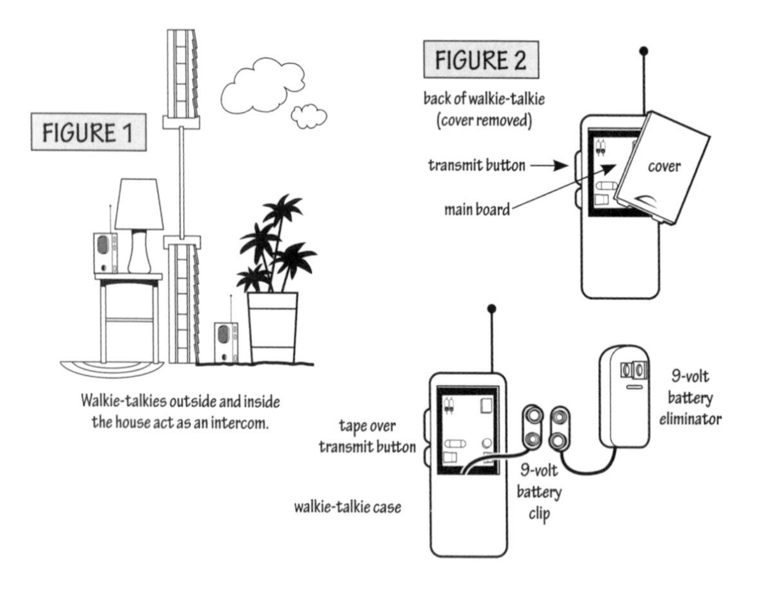

Walkie-talkies outside and inside the house act as an intercom.

For constant monitoring applications, (e.g., as a baby monitor), apply a piece of strong tape across the talk button so the unit is always in transmit mode. This application will eventually drain the battery, so you may want to use walkie-talkies that use 9-volt batteries. Then you can attach an AC battery eliminator to the battery clip (available at electronic parts stores), so it can always be in the on or standby mode without requiring frequent battery replacement. See **Figure 2.**

Sneaky Listener

Use the Sneaky Listener application, similar in operation to the Sneaky Intercom, when you want to monitor a remote location secretly. Simply place one walkie-talkie out of sight, within or under an object. Place the power button in the on position and put tape across the talk button to keep it transmitting.

As shown in **Figure 3**, you can monitor the audio in the area with the other walkie-talkie (and retrieve the remote one later). Or, a walkie-talkie can be placed in a jacket (that's left in a room) to monitor nearby sounds from afar.

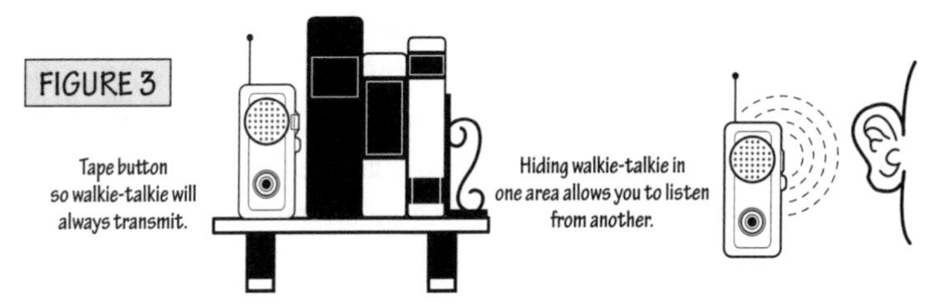

Sneaky Alarm Trigger

A walkie-talkie set provides an inexpensive quick-to-set-up option for a sneaky wireless alarm system. One walkie-talkie set up with tape across its talk button, to keep it in the transmit mode, can broadcast a warning signal to the remote unit. The trick is to place an insulator strip, from a coated business card or a bookmark, between one battery terminal and its clip. Connect a thin strong nylon thread or wire to the other end of the insulator strip and wrap it around the item you want to protect.

FIGURE 4

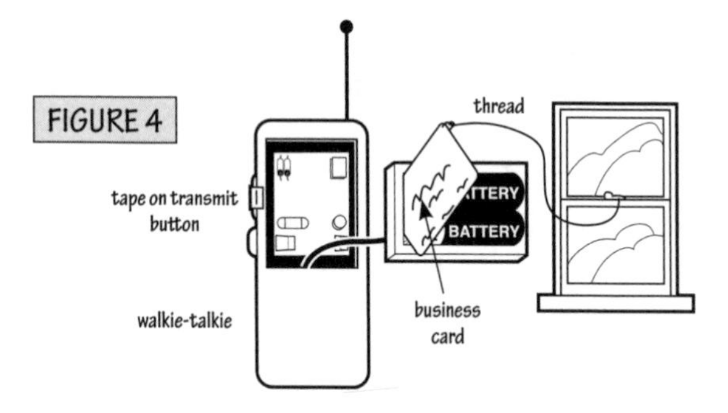

First, open the first walkie-talkie and remove one side of the 9-volt battery clip so that it rests on top of the battery terminal. Then poke a small hole in the insulator strip and tie the thread through it into a knot. Attach the other end to a window handle, door knob, or other object that you want to keep from being moved.

Next, place the insulator strip between the battery terminal and the battery clip. The battery clip should still be attached to the other battery terminal and should keep pressure on the insulator. If the insulator is pulled away from the battery, the battery clip should rest on the top of the battery terminal and turn on the walkie-talkie. If not, wrap a small rubber band around the battery and the battery clip, to apply more pressure. See **Figure 4**.

Turn on the walkie-talkie's power button and tape the transmit or talk button so it stays on. Place it out of sight from the window or object that it will be connected to with the thread.

When someone opens the window, the thread will pull the insulator away from the battery clip, turning on the walkie-talkie, which will transmit to the other walkie-talkie.

Note: If the walkie-talkie includes a signal button or a Morse code signaler, it can be taped in the on position too.

Candy Packaging Tricks

Sneaky scavengers can reuse the innovative extras that candy makers include with their products. Some packages contain light- and sound-producing cell phone toys, spring-loaded containers, tongs that light up when used to grasp gummy candy, and electric fans with light shows. You'll find valuable adaptation ideas below.

What's Needed

- Candy with accessory included

What the Toys Provide

Figure 1 illustrates several common candy packaging devices ready for modification. A candy fan includes a toy motor, battery, and switch, shown in **Figure 1A** and **1B**, that can be used to replace drained batteries or broken motors in other toys or devices.

A spring-loaded candy stick makes a great sneaky security trigger device. After the cover is opened and the button on the side is pressed, the candy stick pops up and out. See **Figure 1C**.

Toy cell phones make great alarm noisemakers as shown in **Figure 1D**. When you open a toy cell phone case with a small screwdriver, you can easily spot the button and contacts on the printed circuit board. If additional wires are wrapped around or taped to the contacts, they will activate the cell phone's light and buzzer, too. See **Figure 1E**.

Candy tongs, shown in **Figure 1F**, light when squeezed. They provide the essential parts to make a sneaky security trigger device.

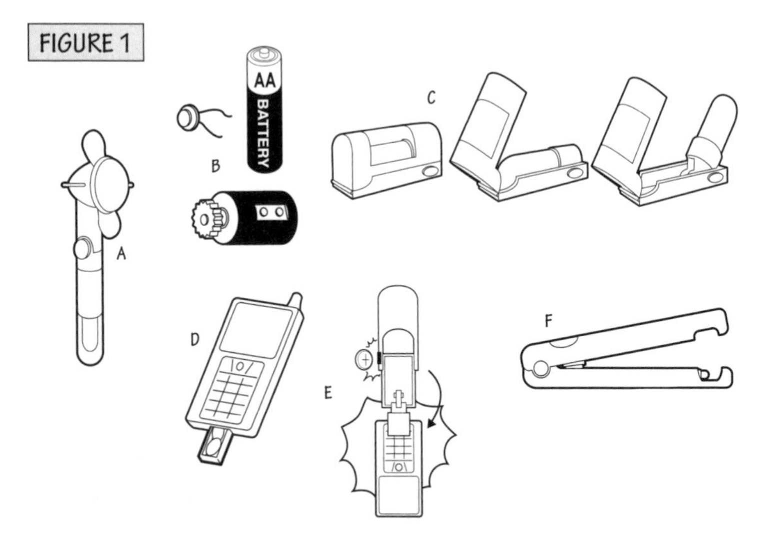

Adapting the Toy Accessories

It's easy to wire the candy tongs to activate the cell phone toy. Remove the switch wires from the candy tongs and connect them to the cell phone's push-button contacts on the PC board to activate the light and sound remotely. See **Figure 2A**. Tape the candy tongs to a door hinge or in the back of a drawer to alert you of a breech.

In another example, the candy stick can trigger the toy cell phone when a door that has a Gcell battery taped to it is opened. See **Figure 2B**.

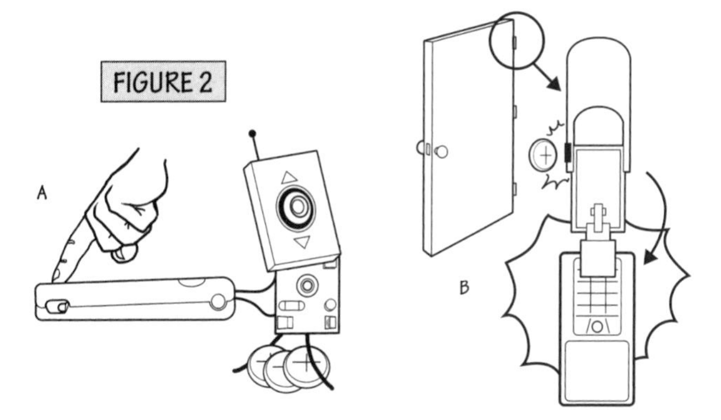

A toy security trigger device can be wired to an existing alarm or connected to a radio-controlled car's transmitter circuit to alert you when a basement floods or when doors and windows are breeched. The alarm receiver can activate a toy cell phone or a similar noise-maker as shown in **Figure 3A**. For example, if you connect two wires across a radio control's activator button, you'll be alerted of an entry breech or when your valuables are being removed. **Figure 3B** shows the wires from the candy prong switch to the transmitter PC board.

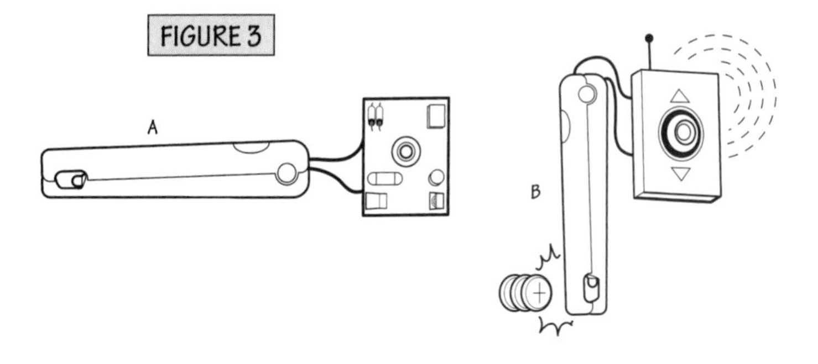

Part IV

Sneaky Animals and Humans

Would you believe a glider was constructed inside a prison using parts from prisoners' beds? Or six lives were saved on September 11, 2001, in a World Trade Tower elevator with just a window washer's squeegee? It's true.

As a primer to being resourceful, you will learn about ingenious improvisers (both human and animal) that use either ordinary items or evolutionary techniques to survive difficult situations.

You will learn about the special World War II British MI9 division that inspired writer Ian Fleming to create the "Q" (Quartermaster) character for the James Bond novels. You will see, with illustrations, how compasses, maps, saws, and other gadgets were hidden in officer uniforms and in packages mailed to prisoners of war. You will also read the incredible story about the Colditz Glider, a 2-man, 19-foot glider that prisoners made using only wood and sheets from their beds!

Did you know that a certain fish shoots at insects with water? Or a spider walks on water and even catches fish? You'll also learn how sneaky creatures hide in plain sight or fool their predators into thinking they are more fearsome than they really are.

This section highlights amazing real-life human and animal survival techniques accomplished with handy items. The lesson? You can do much more than you think with the things around you.

Sneaky Animals

Water Walkers

Molecules on the surface of water pull on each other to create surface tension, which allows some insects and animals to seemingly defy the laws of physics by walking on water.

Water Strider

Just as some people lie on a bed of nails without harm or use snow shoes to prevent from sinking into the snow, the water strider, shown in **Figure 1**, distributes its weight over a large surface by spreading its long back and middle legs over the water's surface. This allows it to walk across the surface of water.

FIGURE 1

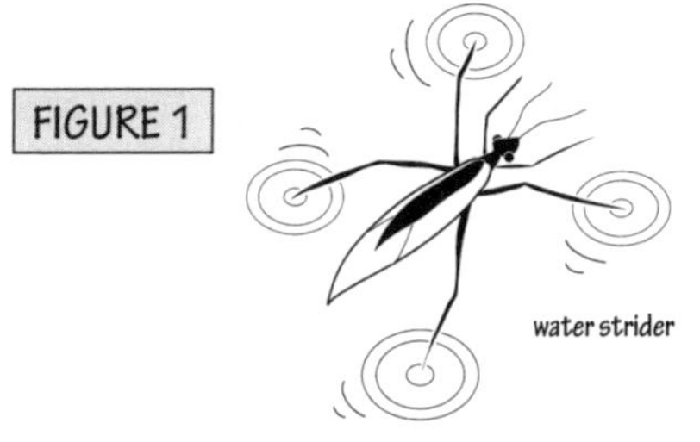

Basilisk Lizard

The basilisk lizard is another creature that can spread its feet, and weight, enough to skim along the water's surface without sinking. Its foot has a broad sole that pushes down on the water, creating an indention like an air bubble in the water. The basilisk lizard pulls its foot upward before the cavity collapses, which reduces the downward

forces and allows the reptile to keep moving. The basilisk lizard must keep its speed high enough otherwise it will sink. See **Figure 2**. Its nickname is the "Jesus lizard," which presumably comes from its ability to walk on water.

Western Grebe

Western grebe birds are among the largest water walkers. During mating season, they perform a courtship display called rushing. After a grebe calls out loudly with a "creet creet" sound, another grebe will answer the call and go through various head bobbing and other motions. Grebes have lobed feet with folds of skin that hang down from their three long toes. The skin folds are flattened to produce a large surface as shown in **Figure 3A**. Eventually the grebes hold their heads up and wings stiff to stride up to 60 feet along the water's surface before diving headfirst into the water. See **Figure 3B**.

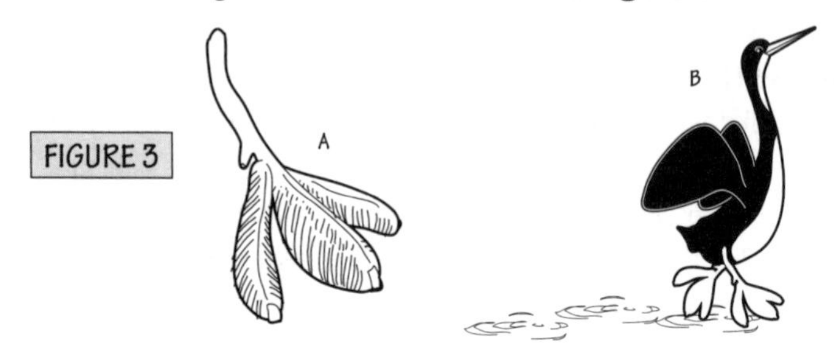

Fisher Spider

The fisher spider strides in a slouched position keeping nearly all of its body and legs on the water's surface. The fisher spider spins a web and attaches it to an object on the water bank. The web controls the fisher spider's movements so it doesn't slide past its prey. It also jiggles the web to attract prey, including insects and fish up to three times its size! The fisher spider can grab a fish with its front legs and sink its fangs in it to inject deadly venom. See **Figure 4**.

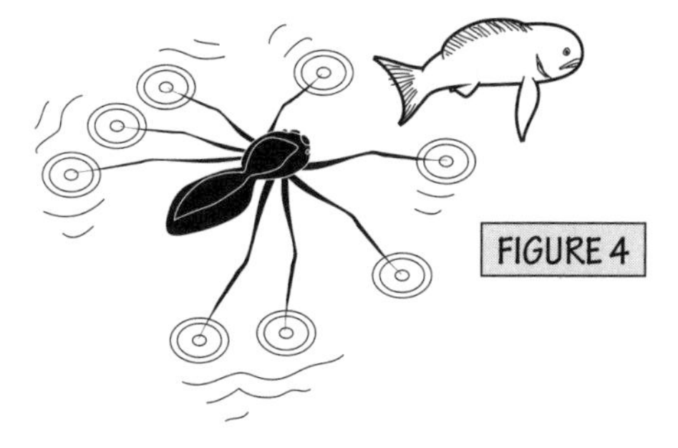

FIGURE 4

Parasite

Cuckoo Bird

The cuckoo bird is a brood parasite, or nest stealer. The female cuckoo bird searches for a host bird that has temporarily left her nest. The cuckoo bird then pushes out one of the host bird's eggs from the nest and lays one of her own eggs there. When the host bird returns, it incubates and cares for the cuckoo bird's egg until it hatches.

The baby cuckoo is also a born sneak. As a newborn parasite that takes advantage of a nonrelative's nest and nurturing care, it usually hatches before the other eggs and pushes the other eggs out of the nest. If the baby cuckoo discovers another newly hatched bird, it will push it out too. See **Figures 5** and **6**.

FIGURE 5

FIGURE 6

Tool Users

FIGURE 7

Japanese Crow

Many birds have been observed using twigs as tools and dropping eggs or nuts from high altitudes to crush them. One of the sneakiest birds, the Japanese crow, has been seen crushing nuts by carefully placing them under automobile tires stopped at intersections. See **Figure 7**.

Archerfish

The archerfish is the only tool user that is not a mammal or a bird. It swims near a bank of swamps or streams just below the water's surface looking for insects on nearby tree branches. When it spots an insect, the archerfish squeezes its gill covers, which forces water through a groove in the roof of its mouth and tongue. The fish uses this water stream as a projectile to disable the insect and knock it into the water for food. See **Figure 8**.

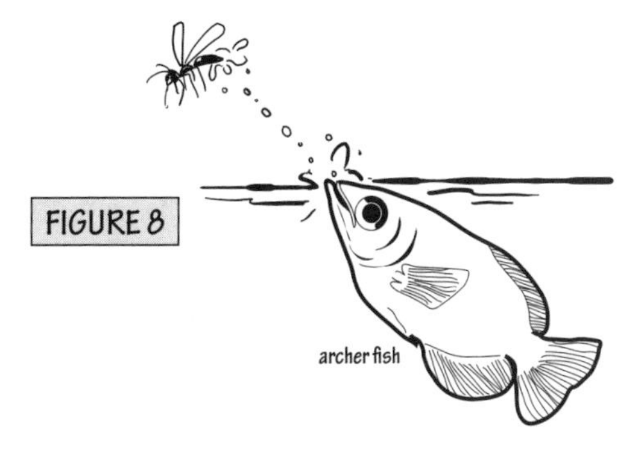

Look-alikes

Some insects and animals use camoflauge as protection. Others avoid becoming prey through spots or feathers that resemble larger creatures, thus confusing or scaring their enemies. In some cases, insects and animals fool predators into thinking their eyes are in another place so if they are attacked, their eyes are protected to prevent blindness and possibly death.

Swallowtail Caterpillar

The swallowtail caterpillar has large spots on its back that appear to be "eyes." When confronted by a predator, the caterpillar turns and swells up the front of its body to reveal large "snake eyes" that scare the attacker away. See **Figure 9.**

FIGURE 9

Caligo Butterfly and Emperor Moth

The caligo butterfly and the emperor moth have large spots on their wings that mimic the appearance of an owl or a large fearsome creature. See **Figure 10.**

FIGURE 10

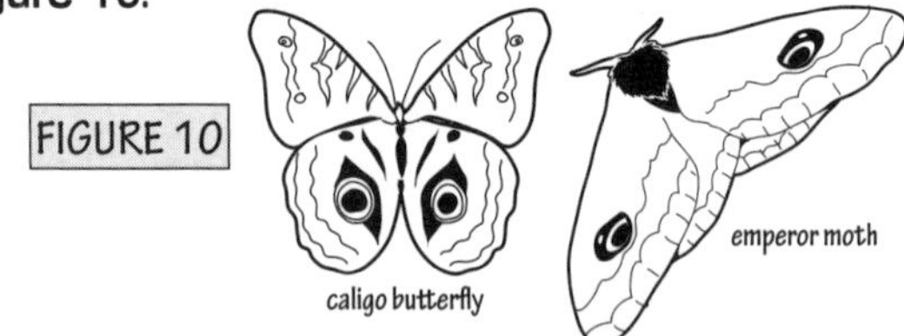

Butterfly Fish

The butterfly fish has two large eye-like spots near its tail that confuse attackers about which end is the head of the fish. The fish's real eyes are disguised in a dark stripe on its face as shown in **Figure 11.**

FIGURE 11

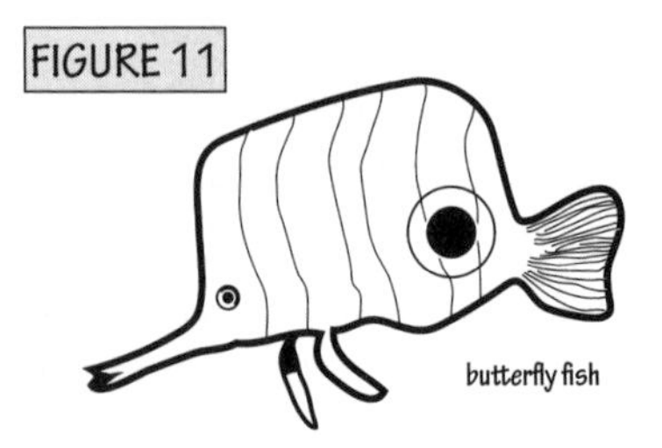

Pearl-Spotted Owl

The pearl-spotted owl confuses its prey and attackers with dark feathers on the back of its head that look like eyes and a beak. See **Figure 12**.

FIGURE 12

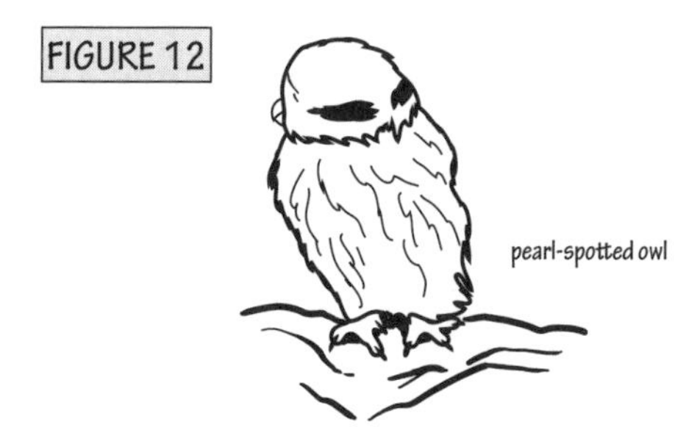

Sneaky Humans

Jan Demczur

On the morning of September 11, 2001, Jan Demczur, a World Trade Center window washer, entered an elevator and was soon trapped with five other men when an American Airlines jet hit the top of the North Tower. Once smoke and heat entered the elevator, the men knew they could not wait for help to arrive and they had to do something immediately to escape.

Jan used his squeegee like a crowbar (see **Figure 1**) to pry open the doors but, unfortunately, they were riding in an express elevator and there was no door on the other side. The men then took turns using the squeegee as a chisel to eventually cut a hole through five layers of drywall. The men climbed through the hole, ran down thirty flights of stairs, and within five minutes the entire building collapsed. Using the squeegee in two sneaky ways allowed them to escape. Jan later donated his lifesaving squeegee to the Smithsonian Institute's September 11 Memorial Exhibit.

FIGURE 1

Colditz Castle Prisoners of War

At a German World War II prison camp called Colditz Castle, British and American prisoners of war built a small workshop hidden behind a fake wall in a chapel attic. Amazingly, the prisoners were able to obtain plans from the prison library to construct a 2-man, 19-foot glider with a 33-foot wingspan. They used wood from their beds and floorboards and cloth from their sheets and sleeping bags to construct the Colditz glider shown in **Figure 2**.

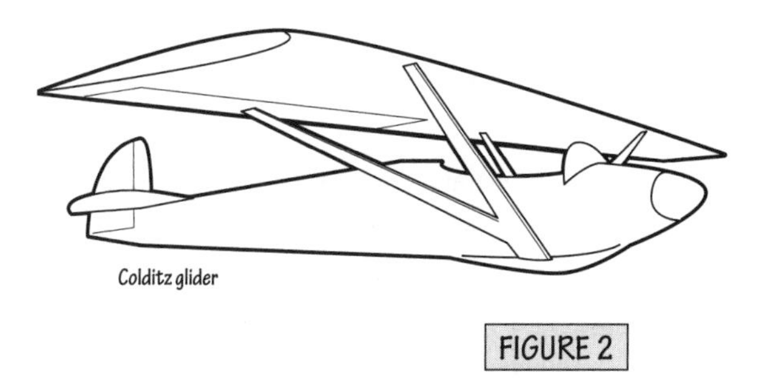

FIGURE 2

Their plan was to place the glider on a roof and to tie it to an iron bathtub with bedsheets. The bathtub was to be positioned on the far end of the roof. When the bathtub fell off the roof, the glider would be pulled across the roof's "runway" and, after disconnecting from the bedsheets with a quick-release mechanism, would fly to freedom.

The prisoners were eventually rescued before they had a chance to fly the glider. However, PBS produced a *NOVA* episode about the Colditz Castle prisoners and the crew rebuilt the glider using the same plans and materials.

At the end of the episode, the host brought some of the surviving British prisoners to an airplane hangar and showed them the replica. They couldn't believe their eyes and some of the men cried on the spot. After the personnel touched and photographed the glider, it was pulled by a pickup truck with a nylon cord and, after a few moments, the glider soared and flew on its own for fifteen minutes!

World War II British Royal Air Force

The British MI9 division (Military Intelligence Section 9) outfitted their World War II Royal Air Force pilots and captured prisoners of war with various hidden devices. Here are some examples:

- The cockpits in many British airplanes were not heated and pilots were required to wear knee-length boots. Unfortunately, if the plane were shot down, the British pilot would be easily identified because of his unique, tall boots. MI9 supplied the British pilots with fur-lined flying boots that had soft leather around the ankle sections. A small knife was included that allowed the pilot to cut away the top sections and, when combined, the extra leather converted into a shepherd's waistcoat.
- Some Royal Air Force bootlaces and cap badges contained a thin, flexible wire saw that could cut through prison bars.

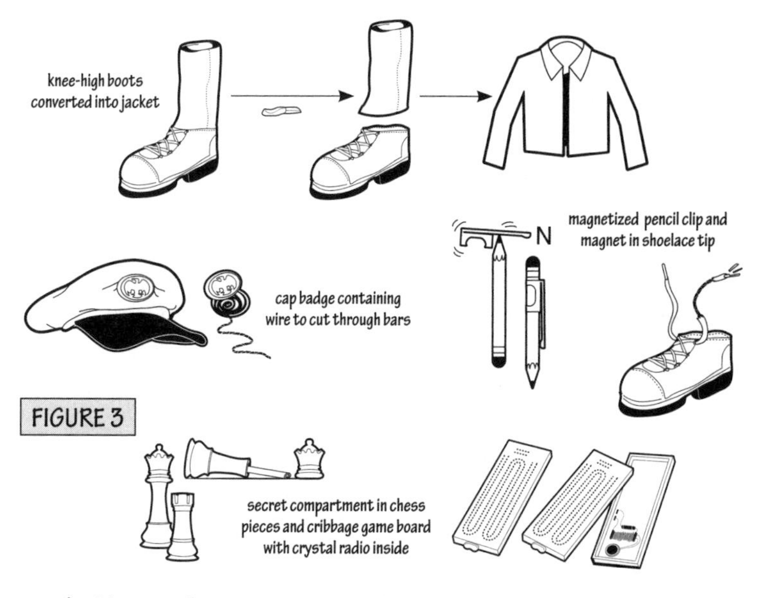

- Maps and compasses were important tools used to complete a successful postescape journey to a neutral nation like Switzerland. Military personnel were supplied with magnetized pencil clips and metal pins inside matches to assist in navigating prison escapes. Small magnets and maps were hidden inside pens and shaving brushes. Tiny magnets were even installed inside shoelaces tips. Sometimes the shoelace fabric also included a wire saw sewn inside.
- Inmates at some prison camps were mailed items that contained hidden devices such as cribbage game boards with crystal radios inside, silk handkerchiefs with maps printed in invisible ink, and pipes and chess pieces with compartments that contained clothing dye and small explosive charges.

- During their imprisonment, some prisoners used what they had around them to their best advantage. They created crude crystal radio sets with spare wire, pencil lead, and razor blades. They also converted boot heels into rubber stamps to forge documents and used wood from their beds to make crude printing presses and sewing machines.

See **Figures 3** and **4**.

Prison Escapes

Many clever prison escapes have been accomplished using ordinary objects, including:

- Braided dental floss used as a rope. Another prisoner sawed through his cell bars with toothpaste and floss. See **Figure 5**.
- A green felt-tip pen used to color a prison uniform green. The convict walked out of the prison disguised as a hospital orderly. See **Figure 6**.
- A Monopoly game's wheelbarrow piece used to unscrew a heating duct cover. See **Figure 7**.

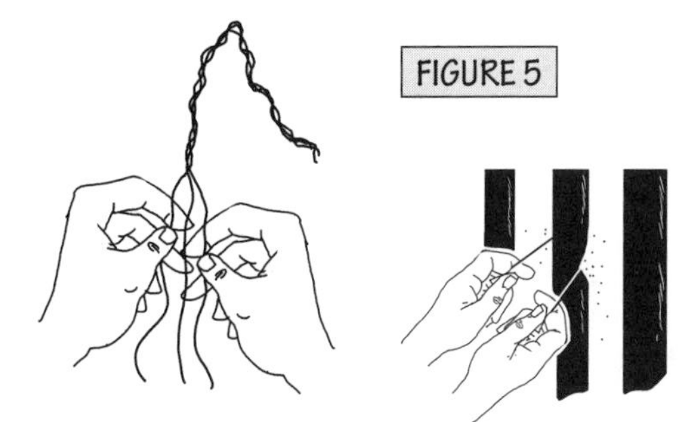

FIGURE 5

FIGURE 6

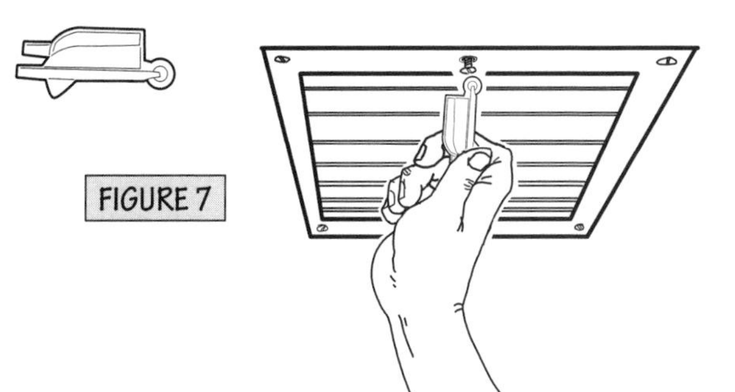

FIGURE 7

Science and Technology Resources

If you find that you like science and invention and you want to go further in your quest for knowledge, this section provides a multitude of science education resources. You'll find links to science fairs, science camps and schools, science organizations, and educational scholarships. You'll also find special inventor resources and contests, grants and awards, free government programs, educator lesson plans, and additional links to free science project Web sites.

Science Freebies, Grants, and Scholarships

www.science.doe.gov/grants
www.siemens-foundation.org
www.thehomeschoolmom.com/teacherslounge/freebies.php

Inventors and Inventing

www.inventorshq.com/just%20for%20kids.htm
www.inventorsdigest.com
www.inventivekids.com
www.super-science-fair-projects.com
www.funology.com

Science Fairs

physics.usc.edu/ScienceFairs
www.astc.org/sciencecenters/find_scicenter.htm
www.childrensmuseums.org/visit/reciprocal.htm
www.campresource.com

Science Sites

www.kidsinvent.org
www.wildplanet.com
www.build-it-yourself.com
www.sciencetoymaker.org
theteachersguide.com/QuickScienceActivities.html

Gadget Sites

www.popgadget.net
www.gizmodo.com
www.thinkgeek.com
www.johnson-smith.com
www.scientificsonline.com
www.smartplanet.net

Science and Technology Sites

www.howstuffworks.com
www.midnightscience.com
www.scitoys.com
www.scientificsonline.com
www.scienceproject.com
www.about.com

Recommended Books

Kids Inventing! (Susan Casey, Wiley)
Gonzo Gizmos (Simon Field, Chicago Review Press)
Joey Green's Encyclopedia of Offbeat Uses For Brand-Name Products (Joey Green, Prentice Hall)
Great Inventions That Have Changed Our Lives (Ira Flatow, Perennial)

Magazines

Make
Boys' Life
Craft
Popular Science
Popular Mechanics
Nuts and Volts